Aviation In[illegible] Directory

2001

AVIATION INTERNET DIRECTORY

A GUIDE TO 500 BEST AVIATION WEB SITES

By John A. Merry

McGraw-Hill

New York Chicago San Francisco Lisbon London Madrid
Mexico City Milan New Delhi San Juan Seoul
Singapore Sydney Toronto

Cataloging-in-Publication Data is on file with the Library of Congress

McGraw-Hill

A Division of The ***McGraw-Hill*** *Companies*

1 2 3 4 5 6 7 8 9 0 DOC/DOC 0 9 8 7 6 5 4 3 2 1

ISBN 0-07-137216-4

The sponsoring editor for this book was Shelley Ingram Carr, the editing supervisor was Steven Melvin, and the production supervisor was Pamela Pelton. It was set in Berkley by Joanne Morbit of McGraw-Hill's Professional Book Group composition unit, Hightstown, N.J.

R. R. Donnelley & Sons Company was printer and binder.

This book is printed on recycled, acid-free paper containing a minimum of 50% recycled, de-inked fiber.

To my mother.
She continues to read every word of these books
even though she doesn't fly and rarely surfs the Web.
Now that's love.

Contents

Aviation Directories

Weather

Pilot Resources

Flight Training and Flight Schools

Aviation Online Magazines and News

Aviation Parts, Supplies, and Aircraft

Aviation Entertainment

Introduction

It still amazes some people to think that a book containing aviation's best Web sites would be of interest to online aviation surfers, much less the overwhelming best-seller it has become. "It's better to just surf the Web with a search engine or two," they suggest. But the fact remains that searching the Web is time consuming at best. Even with today's search engine technology and some knowledge of complex searching, you'll still be wading through time-wasting sites. Trust me. Couple the time-saving factor with the sheer convenience of earmarking sites of interest while not being parked in front of a monitor for hours, and you've got quite a resource. A best-seller even.

Aviation Internet Directory: A Guide to 500 Best Aviation Web Sites is the third edition to the family of *Best Aviation Web Sites* books, resulting from thousands of hours of searching, clicking, and reviewing. Here you'll find award-winning sites ranging from aviation news to pilot resources, hand-picked by pilots for pilots.

Besides the addition of 200 carefully chosen aviation Web sites, *Aviation Internet Directory: A Guide to 500 Best Aviation Web Sites* also includes updated reviews from the last edition, e-mail addresses, and *free* online updates. As in prior editions, the book's purpose is to help you avoid a time-consuming tangle enroute to aviation's better Web sites. And, rest assured that the choosing was performed in a completely unbiased way. Nothing listed here appeared as a result of paid advertising or other favorable treatment. If we thought a site to be worthy of your time, we included it.

Review Rating Criteria

As with most reviews, the author's subjectivity ultimately becomes the predominant rating criterion. However, knowing that this wouldn't fly with most aviation enthusiasts, I've established a few more tangible guidelines against which *Aviation Internet Directory: A Guide to 500 Best Aviation Web Sites* were judged:

Content: Did I uncover practical data and substance or a cesspool of typos and blurry plane pictures?

Layout/Design: Was I bored to tearful yawns or mystically enthralled with site aesthetics?

Functionality: Is site navigation a frustrating maze of futility or a wondrous example of efficiency?

Overall Audience: Does the site offer benefits to 12 people or a million and twelve?

Scale:

✈ ✈ ✈ ✈ ✈

noteworthy bookmarkable cyber-brilliance

Online Updates

While each site's address and content have been checked (and rechecked), please bear in mind that addresses and page information may change or evaporate completely with time. It's simply the nature of the beast.

To keep up with aviation's dynamic sites, however, you are invited to check into *Aviation Internet Directory: A Guide to 500 Best Aviation Web Sites*' Online Updates page for the latest in additions, deletions, and address (URL) changes. Stop in by pointing your browser to

http://www.bestaviationwebsites.com/bookupdates

Each of the 500 sites found in the book, including the 150 bookmarkable listings, will be monitored continuously for changes and reported to you via the Online Updates page. Whether you're having trouble accessing a site or just want a list of site modifications, give the page an occasional visit. It may save you some frustration.

The Basics of Browsing

Web browsing continues to be one of those endeavors that prompts a spectrum of emotions, from utter frustration to complete elation. It's really an inexact science that defines the word *dynamic*. Slow connections, DSL (high-speed digital subscriber line), cable modem, online traffic congestion, and wireless surfing add to the mix of the good, the bad, and the ugly.

Whether you're new to the Web or surfer savvy, I can't stress enough the importance of using the most current version of your browser (see definition below). Today, Web design gurus pour over HTML code and graphics hoping to serve up a cutting edge experience for you—the aviation surfer. However, if you're flying under the hood with a rusty old browser and a slow connection, you probably won't see the visual treats or hear the roar of a biplane. In short, you'll miss out. My advice is to stay current by downloading the newest browser version on occasion. How do you know when a new version is out? Keep in touch with your browser's home page (*http://www.netscape.com*, *http://www.explorer.com*, etc.) or visit a browser news site such as Browser Watch (*http://browserwatch.internet.com*). Then, simply download the newest version for free. Similarly, plug-ins (see definition below) and helper applications change too. Be sure to have the latest and greatest or prepare to forgo the true multimedia experience.

Although the following may be old news for seasoned surfers, new Internet users may appreciate a quick introduction to Web browsing basics.

First, some terminology:

Browser A software program used to view and navigate Web pages and other information. The most popular browsers include Netscape Navigator and Microsoft Explorer. We recommend using the latest versions of either of these two browsers, since most Web sites are formatted specifically for them.

URL A Uniform Resource Locator (URL), also referred to as a Web site address, points to a specific bit of information on the Internet. For example, http://www.bestaviationwebsites.com is a URL.

Bookmarks/favorites Most browsers offer a convenient way of storing and organizing your favorite Web sites in the form of a "bookmark" or "favorites" index. Adding a site to a bookmark/favorites list saves you the effort of retyping the site's URL during future visits.

Plug-In As you access some sites in this book, you may notice the need for plug-ins, or helper applications, to run video or audio features on the site. When using the latest versions of Netscape Navigator or Microsoft Explorer, you'll discover that most plug-ins are already installed. However, if you find that you're lacking a specific plug-in, chances are you'll be able to download it free. Most Web sites will provide a link back to the plug-in download site. Examples: Shockwave and/or FLASH, RealAudio and RealPlayer, Quick Time, and Adobe Acrobat.

Second, armed with this knowledge, you're ready to begin browsing. Simply type in the site address into your browser's location/address box and press "Enter." Be sure to pay particular attention to any special characters or upper case letters in the URL. To access each site, you'll need to type the address *exactly* as it appears in this book.

Third, be patient. Unforeseen forces sometimes determine your success at bringing up a site. It may be down temporarily. The actual communication lines may be jammed with too many users during peak times. Or the site simply has withered away. Our advice? Try retyping the address another day, and move on to a site that does work.

So go on now, type, click, and bookmark.

Aviation Internet Directory

Airlines

AIRodyssey

http://airodyssey.tripod.com

e-mail: airodyssey@bigfoot.com

RATING

BRIEFING:

First-rate personal collection of commercial aviation goodies.

My goodness, Sergio, you must have spent the better part of your downtime coding, organizing, and writing the commercial aviation catch-all site of AIRodyssey. That doesn't include the countless hours keeping the thing updated. For those of us intrigued by the commercial "big boys," your efforts are appreciated.

Top-site, color-coded tabs direct you to your destination quickly: articles (lots!), quizzes (updated every 2 weeks), references (airline and airport codes, radio communications, airline logotype museum, call signs, and more), and departures (free e-mail, flight tracker, and more). No, there's not much in the way of "million dollar" site design, but I'll take this clean, easily navigable, penny-pinching version anytime.

Frankly, I found the Articles area to be best. Look for lots of good, quality information from a variety of topics, such as airports, air safety, flight stories, media, society and humor, then and now, and travel issues.

Free or Fee: Free.

Airliners.de

http://www.airliners.de
e-mail: dhasse@airliners.de

RATING

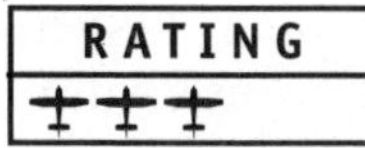

BRIEFING:

Vast collection of civil airplane knowledge spanning the time of the first jet engine to today.

Translated from German to English for us English-only folks, Airliners.de proves itself to be *the* source for civil airplane information "from the time of the first jet engines until today." Whether it's the German or English version for you, you'll appreciate nice organization with frames, solid site navigation, and excellent content.

Yes, as you might expect, there are airliner pictures, enough to make you cringe at the clock. But as the site developers mention, there's more. Much more. Take a peek into the world's aircraft data and history, aviation terms, reader forum (mostly German contributions), links, of course, and plenty of flight simulator downloads and information.

One of my favorites? Sure there are many. But be sure on your next visit to browse the enormous aviation "dictionary." Prepare for inundation of knowledge.

Free or Fee: Free.

AirlineQuality

http://www.airlinequality.com
e-mail: info@airlinequality.com

RATING
✈✈✈

BRIEFING:
No airline schedules or promo stuff here, just brutally honest rankings of the world's airlines.

Pitted as "the world's first global quality ranking system for airline product & service standards," AirlineQuality does the grueling, number-crunching dirty work and tallies, catalogs, and maintains results from customer surveys.

Let me be clear: This isn't some frilly, promotional, paid-by-the-airlines kind of site. The unbiased information comes from real travelers like you and me. Although AirlineQuality doesn't provide airline schedules, frequent-flyer programs, or the like, it does serve up current stats (rated with stars) and facts about actual standards of service for each airline.

Presented simply, yet professionally, your left-frame menu guides you into information on the star ranking, reports, surveys, news pages, FAQs, and more. Once you delve into ratings for individual airlines, you'll get much more detail. Expect all the specific star ratings for each airline's overall class rankings and in-flight features (seating, safety, cabin staff service, catering, entertainment, and more).

Look to AirlineQuality for an eye-opening perspective into the opinions of many.

Free or Fee: Free.

Cruisinaltitude.com

http://www.cruisinaltitude.com

e-mail: mskonfa@yahoo.com

RATING

BRIEFING:

Not much to do around the house? Weather's got you grounded? Click in for a fun time-waster.

Conceived at 37,000 feet over southern Paris on the back of a piece of paper with the map of France on it, Cruisinaltitude.com is not your prototypical commercial aviation enthusiast site. Its contents spill over into uncharted territory.

For those who just seem to tingle when the words *commercial aviation* are uttered, this site and its strange wonders cater to you. True travelers and enthusiasts will enjoy a wide variety of photos and seat maps (of specific carriers), special-events listings (new aircraft rollouts, etc.), airline schedules (courtesy of FlightLookup.com), and tail photos (complete with location, airline, date, aircraft type, and more). Hopefully, you're starting to get the idea. It's commercial aviation euphoria plain and simple.

Step into the Trip Log for a fun little comparison of traveler mileage. Quite a few travelers continue to document their journeys with this online database (name, location, miles flown, countries, airlines, favorite airline, etc.). Logging your trips is free and relatively easy. Go ahead, see where you rank. You might even find your way into the prestigious Trip Log Hall of Fame. Search the log by miles flown, airports visited, airlines flown, countries visited, longest segments, shortest segments, and yes, the "wackiest trip."

Free or Fee: Free.

FrequentFlier.com

http://www.frequentflier.com

e-mail: webmaster@frequentflier.com

RATING

BRIEFING:

Sure, the world of frequent-flyer programs is complicated, yet booming. Stop in here for enlightenment.

Prior to describing FrequentFlier.com's massive stockpile of content, it should be said that finding your way throughout this site is painless. Thanks to good text link menus and description combinations, you'll begin to fly around the complex maze that is the frequent-flier program (FFP). Pull-down menus and fancy graphics also provide a beacon in the murk.

Content, however, is king. Lots of information, opinions, suggestions, facts, figures, and contact numbers fill the screen with every click. There's much to learn: How to choose an FFP, how to earn the most FFP miles, how to get the most "return" from those miles, and what to do about expiring miles (includes a cool real-time countdown to the annual expiration deadline).

Actual intro page menu choices include Miles from Credit Cards, Frequent Flyer Programs (links and advice), The FrequentFlier Forum, and The FrequentFlier Polls. Yes, there's more. But add your bookmark now and peruse the whole thing later.

Free or Fee: Free.

World-Airliners.com

http://www.world-airliners.com
e-mail: online form

RATING
✈✈✈

BRIEFING:

A breath of fresh air in commercial aviation enthusiasm. Photos, message board, and more are standing by.

I always find it refreshing to uncover private, noncommercialized, enthusiast-driven Web sites among the dot-com craze. Simple, pleasantly presented sites provide a nice distraction from the money-hungry world of Web commerce. Enter World-Airliners.com.

Professionally designed to be easy to use and simple to navigate, World-Airliners.com would simply "like to share its interest in civil aviation with friends all over the world." That's it. End of story. The main feature is really the Photo Gallery, where photos from photographers all over the world are displayed. My count puts it at 1800 or so at review time. However, if you jump in a little deeper, you'll find that there are even more reasons for a bookmark. A message board shares the opinions of many worldwide enthusiasts spanning countless topics. Airlines reviews, aircraft reviews, and airline news pretty much round out the main contents of the site. Aircraft model stuff finds its way into the mix, too, if you're into that sort of thing.

Free or Fee: Free.

AirportHub.com

http://www.airporthub.com
e-mail: info@airporthub.com

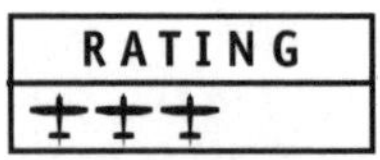

BRIEFING:

This "global airport portal" serves as the perfect stopover on your way to worldwide airport information.

Kind of the Chicago O'Hare International of airport portals, AirportHub.com envelops the airport world with its broad offerings. Passengers, crew, airport administrators, and businesses involved with airports may want to clear some space for a bookmark right now. This one's a keeper.

Overall design, layout, and site navigation work around a lightening-fast database-driven engine. Almost entirely text-based, menus and information links are perfectly presented and quick to load. Specifically, news, articles, press releases, jobs, requests for proposal (RPs), links, and airport information are neatly tucked away in menus, ready for the clicking. All are current. All are worthwhile.

As an added bonus, AirportHub.com eRFP and Auction Services offer real-time bidding solutions for buyers and sellers of airport-related products and services. These excellent online services help match the needs and solutions of those specifically focused on the airport industry.

Free or Fee: Free.

Airliners.Net

http://www.airliners.net
e-mail: admin@airliners.net

BRIEFING:

So you enjoy airliner photos? Try over 15,000 with Airliners.Net.

The jumbo 747 overseeing your browsing options gives you perfect symbolism. Airliners.Net's photographic database and entertaining feature selections are enormous—almost exceeding gross takeoff weight. My browser got a bit bogged down, but loading time is worth it. Really. Once you're into and out of ground effect, the spectacular view takes shape.

Your first visual references begin with page design and organization. From the graphically spectacular main menu, site options are thoroughly summarized. Instant photo search capabilities, with basic or advanced options, use simple category selectors and pull-down lists from which to choose. Side text menus even line the page perimeter with unlimited airliner factoids and picture options.

Next, begin sampling the nice variety of pages. Aircraft Stats & History, Aviation Forum, and Live Aviation Chat are a good start. Move even further into the Lighter Side (funny aviation pictures with commentary) and a great aviation store for more airliner extras.

Lastly, the real behemoth may be found in Airliners.Net's inexhaustible database of over 100,000 photos. Yes 100,000. Give your search criteria or choose from many predefined sections: Our Best Shots, Classic Airliners, Military Aircraft, Special Paint Schemes, Accidents, Airport Overviews, Flight Decks, Air to Air, Aircraft Cabins, and more. Many more.

Free or Fee: Free. Just remember to register (also free). You'll see a personalized front page, newly added photos since your last visit, and e-mail notification of changes.

Swissair

http://www.swissair.com
e-mail: online form

RATING

BRIEFING:

This brilliant dissemination of Swissair information is ripe for a bookmark.

First, a few of my Swiss favorites: chocolate, watches, skiing, and now Web perfection in the form of Swissair's cyber presence. Web-wise, Swissair has incorporated all my page-design favorites juxtaposed with a team of navigational essentials. Text links, site map, and pull-down "pick" boxes join forces with a summarized menu of commonly used functions to tantalize world travelers with Internet bliss.

Content-wise, Swissair's travel planner is especially worthy of note. With its city tips and thoughtful destination data bank, you become distinctly aware that Swissair is a prominent world carrier for a reason. Read on about such places as Amsterdam, Athens, Boston, Budapest, Hamburg, Lisbon, Osaka, Vienna, Zurich, and others. Even topics such as Swiss tourism, exchange rates, railways, special interests (golf, youth, and families), and airport maps reward the viewer with useful and detailed insight.

Of course, if you're more in the mood to skip the fluff, just go straight to Booking by Flights (thoughtful instruction provided if you need it), schedule information, arrival/departure information, or Qualiflyer personal account data.

So easy. So thorough. So convenient. So Swiss.

Free or Fee: Free.

Frontier Airlines

http://www.flyfrontier.com

e-mail: online form

BRIEFING:

Destination? The new Frontier.

Cyberspace—the new frontier for Frontier. Don't be confused. I'm not exploring a historical tribute to the Frontier of old—the airline that ceased operations in 1986 following its acquisition by Newark-based People Express. I'm inviting you to dabble online into today's Frontier Airlines, which lifted off again in 1994.

With its Denver hub, Frontier Airlines shuttles folks east and west via wildlife-laden 737s and a clickable mouse. The latter's a bit less expensive, but its information is rich. A simple top-margin menu tells the story at a glance: hometown hospitality and straightforward style. Your simple selection guides you through the site's pages of Specials, Programs & Services, News, About Frontier, Employment, FAQs, and Site Map.

The best feature? An always-handy schedules, fares, and reservations request form in the bottom frame. It's easy to use and almost too convenient.

Free or Fee: Free.

Hawaiian Airlines

http://www.hawaiianair.com

e-mail: online form

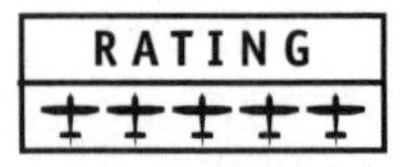

BRIEFING:

A taste of tropical pleasures awaits the online browser here.

Yes, I became tempted immediately. As soon as I reviewed the opening page's current Honolulu weather conditions (75.9°F), I was just about to book online. Reviewing a site like Hawaiian Airlines is probably not the best idea during a bitter cold winter morning. My weather woes aside, Hawaiian Airlines' site will warm anyone's online experience year-round.

Selection and style rule the Web wonders of the "Wings of the Islands." On loading, your first stop succinctly provides you with a summarized view of a dizzying array of options. Quick Flight Reservations prompts you through a few quick questions to begin your journey. Flight Schedules provides flight information for the six beautiful islands of Oahu, Hawaii, Kauai, Maui, Molokai, and Lanai (the route map clearly breaks down the options and whets your appetite to get away). Destination information refreshes your memory of the pleasures of tropical bliss. In-Flight Services makes known Hawaiian's stellar in-air services.

Another convenient nicety is the listing of all pertinent company contact information at the bottom: corporate address, mailing address, phone, fax, e-mail, and special reservations phone numbers (if online booking's not for you).

Free or Fee: Free.

Midwest Express Airlines

http://www.midwestexpress.com
e-mail: online form

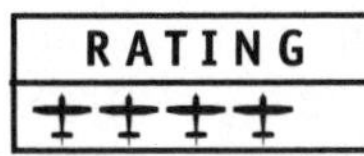

BRIEFING:

A gutsy little airline that continues to win awards in "real life" and now online.

No, the company is not a globally minded airline powerhouse. However, you'd never know it from the company's respectable Web site offerings. With flavor and flair, Midwest Express Airlines makes its mark online with a surfer-friendly site. This Milwaukee-hub airline packs a resourceful punch with a barrage of electronic pages. Ease your seat back and join me for a recap.

Midwest Express carefully makes use of all your favorite navigation essentials: fancy buttons, simple text menus, quick searching, a thorough site map, and embedded description links. However, site mechanics alone does not earn mastery. The real mastery found here may be in the site's comprehensive information offerings. Get flight information (routes, schedules, destination information, fare quotes, vacation packages, and more), learn about the company, and check into frequent-flyer information (current programs and promotions, member account information, enrollment, redeeming miles, and more). Or be reassured with thorough descriptions of outstanding services (dining, industry awards, maintenance and safety, aircraft statistics, seat maps, and more).

Clever little extras also greet you on startup, such as a quick flight status lookup, a local cities weather finder, and a worthwhile private tour of the company's MD-80 aircraft.

Free or Fee: Free.

Air France

http://www.airfrance.com
e-mail: online form

BRIEFING:

Parlez-vous Français? Doesn't matter. English and French versions of Air France's site are accessible to all.

Finally, a breath of fresh cyber air, Air France's megasite depository includes much more than the standard airline essentials. Yes, within its cool, European Web shell beats a more entertaining array of variety.

First, a confirmation on what you'd expect. Digesting schedules, booking, frequent-flyer information, travel services, destinations, and Paris hub information fill the better part of a day. However, the beauty (besides subtly striking page design) lies in the extras. For example, you rarely find such a fascinating historical journey of an airline. And Air France's will exceed your expectations. Turn the pages through more than 60 years of history, featuring an evolving fleet and emerging passenger service.

Free or Fee: Free.

Continental Airlines On-Line

http://www.flycontinental.com
e-mail: online form

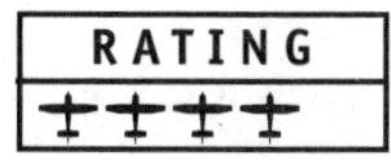

BRIEFING:

Heavily "menu-ized," continental-sized compilation of company offerings.

Nearing the point of overorganization, Continental's site gives you more menu options per square inch than any of its competitors. The daily updated Internet interface sports top and bottom menus, "quick click" topic menus in the body, and icon-oriented button menus. Thus, if you find yourself lost at Continental Airlines On-Line, you probably shouldn't be handling a mouse in the first place. There's just no way to lose your way.

Deals fill up the vacations and specials screen, as do a megasite load of Continental information and options everywhere else. If you're like me and you prefer mini-summaries before you click, try the site index. It's excellent. Then get off the ground with resources such as current flight status and gate, ticket purchasing online, OnePass frequent-flyer promotions, products, and services descriptions, Continental vacations packages, SkyMall catalog shopping, in-flight services, and a healthy list of travel resources.

Oh, and don't leave without at least a click into the Featured Destinations. It's pretty thorough.

Free or Fee: Free.

SkyWest

http://www.skywest.com
e-mail: info@skywest.com

BRIEFING:

High-profile regional commuter arrives on the Web scene in style.

Over 800 daily flights to 65 cities in 13 western states and Canada. Now, you may add to the stats thousands of monthly "hits" to the company's SkyWest Web stopover. Popularity and name recognition are not strangers to SkyWest, even online. A key player as a regional connection airline for Delta and United, SkyWest dazzles Web travelers too with a site worthy of respect from the airline megasite crowd.

Design is stellar, if not cutting-edge. Menus and frames work with dexterity together. Graphics, photos, and maps are easy to read. And helpful contact information is always within a click or two.

Whether browsing as a customer or more on a corporate level, SkyWest serves up equal portions of "five plane" organization and quality information. For example, dipping into the customer-oriented pot of options, you'll find flight reservations (links to Delta or United Online reservations), frequent-flyer program information, a route map, airport layout maps (Los Angeles, Portland, Seattle-Tacoma, San Francisco, and Salt Lake City), aircraft information (photos, QuickTime VR tours, technical information, seating charts, etc.), air cargo information, and customer service.

Free or Fee: Free.

Alaska Airlines/Horizon Air

http://www.alaska-air.com

e-mail: online form

BRIEFING:

Fully functional airline site with full-service flair.

Functional, practical, and proficient seem to best describe not only the Alaska/Horizon Air Web site but also the entire operation in general. The company obviously has maintained its focus on efficiency and service.

A simple, yet well-designed main page flashes your site choices in a no-nonsense manner. Quite frankly, this site's built for speed. Mach II type speed. Top and bottom menu links keep you comfortably focused, while main body information fills the middle.

Currently serving cities in Alaska, Arizona, California, Idaho, Montana, Nevada, Oregon, Washington, Canada, and Mexico, Alaska Airlines/Horizon Air satisfies your curiosity in travel information and trip planning with Web prowess. Flight schedules, a downloadable timetable, online reservations, and flight status information probably will be your most popular choices. However, a huge section on mileage plan information may pique your curiosity. There's also a helpful trip planner, filled with information on travel resources, destinations, a route map, and Q&As.

Yes, even this lower-profile, big-hearted company proudly pushes its own Web specials, an e-mail announcement service, vacation packages, and mile deals. Click in and be pleasantly surprised.

Free or Fee: Free.

Northwest Airlines WorldWeb (NWA)

http://www.nwa.com

e-mail: online form

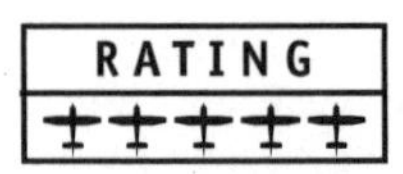

BRIEFING:

Northwest serves up a visual feast coupled with flawless navigation.

NWA's organization, style, and overall Web brilliance follow suit with a long list of online airline award winners. The site is structurally perfect, visually stunning, and thoroughly useful. NWA's WorldWeb encompasses everything the Web is supposed to do: inform, sell, and entertain.

Before I launch into the treasure chest of content, NWA's navigation demands some praise. While I've found most airline sites to be fairly clever with navigational elements, I must say that NWA's WorldWeb is flawless. First-, second-, third-, and sometimes even fourth-level menus keep you easily on track without confusion. As you navigate, new pop-up menus will appear, or simple text links let you drill down further into the good stuff.

And there are many pages that have good stuff. A Travel Planner makes the passenger's life easy with online reservations, future schedules, destination information, travel tips, and more. Northwest Services teaches about on-board service information, gift shopping, etc. And Deals and Promotions tempts your thirst for really getting out and about.

Certainly there's much more. Read about everything site-wide in handy, summarized capsules.

Free or Fee: Free.

Trans World Airlines (TWA)

http://www.twa.com
e-mail: online form

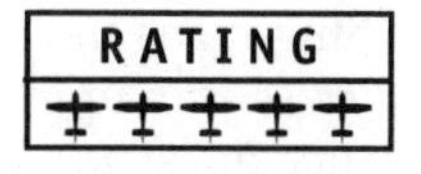

BRIEFING:

TWA—Thoroughly Wondrous Airline site.

TWA, the mammoth of years past and the new vision that soars today, shines in digitized delight online at twa.com. Such a Goliath among airlines, TWA could have opted for megasite photos and larger-than-life graphics. Thankfully, the reins were taken by folks who aim to satisfy the browsing public. Photos, graphics, and menus retain perfectly understated balance. Thus I won't need to summon the bandwidth police. In fact, load time with my average-speed modem was surprisingly fast.

The contents you ask? A well-organized wealth of informational opulence awaits you. Through expandable side menus you'll break into schedules and reservations, frequent-traveler information, information about TWA, getaway vacations, hot deals, airport information, travel agents area, passenger services, cargo services, and an outline of site contents.

Free or Fee: Free.

Air Canada

http://www.aircanada.ca/home.html

e-mail: online form

BRIEFING:

Yes, Air Canada excels with its flashy flair for flight information. But look deeper for some nice extras.

Yet again, when your aviation cyber searching turns up the humdrum, look to high-end airline sites to jazz up the viewing. And Air Canada's cyber presence doesn't disappoint. Rather, this mega-airline stopover gleams with a shiny exterior. Whether you're French-fluent or English-only, the cyber world of Air Canada is instantly at hand. Just point and click.

Wrapped up in a better-than-average package, the traditional airline elements line the pages. Schedules and reservations (cyber ticket office), traveler services (cool travel planning assistant, in-flight special services, in-flight entertainment, etc.), Aeroplan mileage information, vacation packages, and more make planning and preparation with Air Canada simple.

Free or Fee: Free.

America West Airlines

http://www.americawest.com

e-mail: online form

BRIEFING:

Now boarding: efficient online reservations and deals, deals, deals.

Wheeling and dealing online, America West Airlines shows off its power for promotion with a site packed to the rafters with specials, vacation packages, double FlightFund mile deals, and more. Just scan the Highlights. The list of offers becomes so lengthy it's "scrollable." Handy summaries of each whet your penny-pinching appetite, and text links fly you to more complete details.

Once your heart rate recedes from promotion overload, mouse over to the top-margin laundry list of site clickables. Online reservations, products and services, corporate information, vacation packages, travel agent–only information, and a site map tree provide a well-rounded introductory itinerary.

Along with easy-to-read contents, page mechanics fall into the exceptional variety. Main menu icons, text-link backups, and concise summaries steer you properly and avoid the "wild goose chase" approach. Graphics and photos must be lean and mean because loading time is happily quick. The handy site map gives you an easy overhead view of everything. And there's a load to digest.

Free or Fee: Free.

Virgin Atlantic

http://www.virgin-atlantic.com
e-mail: online form

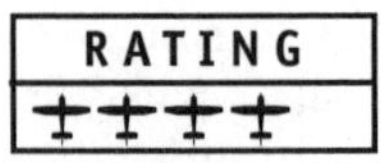

BRIEFING:

Living up to its name for style and service, Virgin's Web site doesn't disappoint.

Custom-tailored to region, your Virgin Atlantic online flight starts out smooth and ends even better. Pick your departure region—United Kingdom and Northern Europe, Greece, India, China, Japan, South Africa, the United States—and move into a host of attractive options. Schedules and booking, services, company news, frequent-flyer information, airport and fleet information, and site search get you started.

More adventurous types will find themselves heading straight for the Vacations area—a whole separate Web site devoted to vacation packages. European vacation packages are many. Just scan the package options: London Breaks, London & Beyond, Fly-Drive Vacations, Land Only, Escorted & Unique Tours, and Specials.

Once you've selected a package and packed your bags, browse through the company's service classes: economy, premium economy, and upper class. Get the details and descriptions for each—even if you're just checking in on the treats of upper class. Hey, economy's just fine too. Find out here.

Free or Fee: Free.

American Airlines

http://www.aa.com
e-mail: online form

BRIEFING:

Find booking and travel information, specials, and overall perfection here.

Only the truly Web competent understand the merging of high-end design with day-to-day practicality. AA.com's Web presentation is so flawless you don't even realize its functionality. It just flows naturally. For example, the introduction page spews organizational perfection with a full list of categorized sections. Each area then has its own pull-down submenu. The best part? Know what you're getting into ahead of time with brief summaries. Even if you get confused, which you won't, members and guests need only visit a well-scripted help area, overflowing with tips and login instruction.

A sneak peek of clickable options includes news at American Airlines, travel planning (fare, gates and times, reservations, schedules), Aadvantage program information, programs and services (too many to list), and corporate information (job opportunities, etc.). Of course, like the others, AA.com promotes its specials. And yes, there's a long list of them. From Today's Specials through Specials Information, this area alone prompts a bookmark.

The key to AA.com's site is becoming an Aadvantage member. With membership, you'll gain access to the heart of the site's wealth: booking flights, making reservations, and managing your Aadvantage account. Sure, on a guest-only basis, there's plenty to see and do. However, become a member for catch-all online participation.

Free or Fee: Free.

Southwest Airlines

http://www.southwest.com

e-mail: none provided

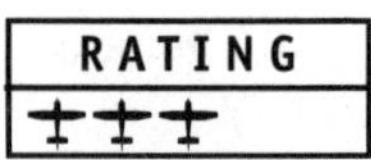

BRIEFING:

How novel—an airline that keeps things simple. It's a great way to fly.

In the spirit of fancy, fun-loving flair, the Southwest Airlines home page sports a happy jet on takeoff—sans the airport chatter and babies wailing. The quick load and variety of "menu-ized" options also will put a smile on your face.

The at-a-glance interface is mostly text-based and easy to use. Steering away from multiple submenus and pull-down lists, the menus point you in the direction of many major site resources. More comes in the form of text links below. Get quick access to schedules, fares, reservations, Travel Center, and Rapid Rewards (frequent-flyer program information and online account services). If you're so inclined, Careers, About SWA, and Programs & Services are also standing by.

Although the mostly text-based method of navigation is kind of plain and simple, practicality prevails with an alphabetized site index. Get a summarized description and clickable links to everything site-wide. From a gallery of video and print ads to city information to shopping with SkyMall, over 60 different site areas are a click away.

Free or Fee: Free.

United Airlines

http://www.ual.com
e-mail: online form

BRIEFING:

All too rare is the combo of great page design and user-friendly organization. Here, all elements are united.

Not one to be outdone by others who spin airline Webs, United checks in with a well-designed (if award-winning) company presence. The company has made excellent use of navigational niceties essential with sites this large. I found myself relying on convenient at-a-glance menus, but a site map, site search, and link menus also guide your way.

Sure, every traveler's whim is answered here. Whether it's reservations and planning, information at the airport, a tour of "in the air" services, Mileage Plus information (FAQs, summary, special offers, partners and program information), or even United's laundry list of services, chances are your questions will be answered online. The Destination Information area gleams with a wealth of informational riches for national as well as international travelers. Get a handy peek into city guides, currency, visa/customs, embassies, foreign languages, baggage information, and weather.

If you're like me, you made the decision to circumvent travel agents long ago and book flights directly online. Today, access to real-time fares and availability information is everywhere. And, of course, United is no exception.

Free or Fee: Free.

British Airways (BA)

http://www.british-airways.com

e-mail: online form

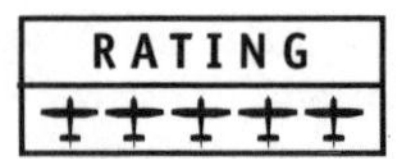

BRIEFING:

Serving up more than just tea and crumpets, BA's online world is entrancing and useful.

You've got to have some pretty high-powered equipment to service the kind of artistry and volume found at British Airways online. The company obviously has invested a bit in design talent as well. For someone like me who spends hours rummaging through hideously designed pages, a layout with real graphic design shines like an airport beacon in the fog. Photo usage, page presentation, database queries, and the like dance in perfect harmony site-wide.

The whole company's represented here one way or another. You get to choose how detailed or simplistic your BA site experience will be. Go quickly to six main topics: Book Online, Special Offers, Flight Info, Executive Club, On Holiday, or Worldwide (pick a BA Web site closer to home). Delve into a world of extraneous yet interesting BA information. For example, special offers, scrolling news and promotions (lots!), company information, great travel Q&As, and more fill your screen.

My favorite site stopover by far involves a tour through personalized travel miniguides. Whether it's Amsterdam, Zurich, or any of the 99 destinations in between, create your own custom-tailored travel guide based on your data input. Select dining, accommodations, language, price range, and sight-seeing preferences. A richly informative miniguide just spits out!

Free or Fee: Free.

Delta Air Lines SkyLinks

http://www.delta.com

e-mail: online form

RATING

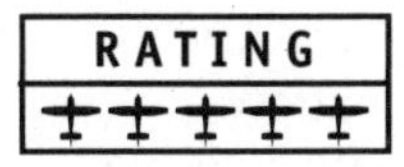

BRIEFING:

Looking for perfection online? Delta's just the ticket.

If it's possible to reach visual and organizational perfection in today's ever-changing Web world, then Delta's done it. Although load time suffers slightly, you'll be happily clicking along nevertheless within the artfully designed realm of Delta Online. Menus, Javascripts, pull-down boxes, help solicitations, and summarized text links form this virtual wonderland. Beauty, brains, and online booking—now that's euphoria.

If you're just browsing, not booking, you could spend forever reading and clicking through every resourceful page. Every nook and cranny offers some special or information tidbit on which to feast. Check out airport layouts to get a leg up on connections and parking. Look through Delta's destination maps. Or find out when Aunt Tootie gets in by checking arrival/departure information. Of course, if you want to skip the glitz and information, jump straight to Reservations, Check Fares, and Buy Online. Oh, and if you're into SkyMiles, one click gets you into the pages to check your current balance. How close are you to earning free travel?

Free or Fee: Free.

US Airways

http://www.usairways.com
e-mail: online form

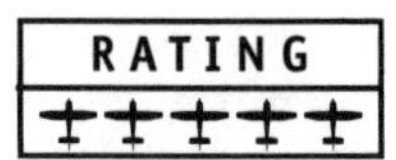

BRIEFING:

An Envoy-class approach to airline Web presentation.

Mirroring the company's style in the air, US Airways shines online too with its catch-all information resource. Clever, clean, and concise, the site gives you the corporate-looking professionalism you'd expect from US Airways. Loading is fast, options are clear, and menus are everywhere.

Your travel choices are many, but not so many that a day of clicking is consumed. Certainly the essentials are at the ready: online reservations booking, schedule viewing, company information, and Dividend Miles stuff. However, as expected, you get a bit more with US Airways. Travel news keeps you current on company news and specials. Car and hotel booking is also conveniently clickable. And extensive passenger information makes research a snap. Read about premium travel, consumer tips (good!), aircraft diagrams, airport maps, baggage requirements, and even the current movie guide for Overture (US Airways in-flight entertainment system).

Another favorite most will appreciate may be found in Promotions. Get up-to-the-minute postings on special offers, special rates, group rates, shuttle fares, and E-savers (special discounted fares via e-mail).

Free or Fee: Free.

Airlines of the Web (AOW)

http://www.flyaow.com
e-mail: online form

BRIEFING:

Fancy airline-only interface combines information, resources, and reservations.

Just step into the well-thought-out terminal of airline online information. Clickable sources for a variety of topics are all within mouse range. But don't worry, the intimidating volume of information behind the scenes is easily accessible with AOW's handsome interface.

Offering way more than just airline information, AOW seeks to become your primary center for reservations, frequent-flyer information, and tips for airline travel. While visiting, however, I gravitated more to the seemingly limitless selection of airlines (actually about 500 worldwide) and the Hangar directory.

To uncover your airline(s) of choice, start by selecting a region. Africa, Asia, Australia, the Caribbean, Europe, the Middle East, North America, or South America. Once you select a region, your airline links unfold—be prepared for a long list. Some links provide handy notes that clue you in about if the site is official or unofficial and its language.

Tired of traversing airline sites? Skip to the aircraft specifications (over 100 aircraft detailed), manufacturers, virtual airlines, former airlines, and more.

Free or Fee: Free.

Cathay Pacific

http://www.cathaypacific.com
e-mail: online form

RATING

BRIEFING:

World-class world carrier knows what it means to be World Wide Web savvy.

I'd say service to 50 cities on five continents thrusts Cathay Pacific into "major carrier" airspace. As such, you'd expect a grand online adventure to whisk you and your mouse away to distant lands. Well, pack your bags before you log on because Cathay Pacific's Web wonders will sweep you up into its network of travel possibilities.

As sure as the world is round, Cathay Pacific casts a brilliant light on true Web style and artistry. The best part? Most who browse the pages won't even realize the stellar design and layout. They'll just happily click through each page with ease and satisfaction, not realizing the underlying brilliance. For example, pages are uncluttered to the point of simple. "Fly-over" menus get you just about anywhere site-wide. And all graphics, illustrations, and photos have been tastefully chosen with quick downloading in mind.

Content-wise, all your air travel expectations will be met with the standard fare of flight information, frequent-flyer propaganda, cyber offers, destination information, aircraft seating and in-flight services, and more.

Free or Fee: Free.

Bookmarkable Listings

Airways Magazine
http://www.airwaysmag.com
e-mail: airways@nidlink.com
Current airline news, commercial air transport–related gifts, and printed magazine subscriptions available online.

Airline Information On-Line
http://www.iecc.com/airline
e-mail: airinfo@iecc.com
Answers to frequently asked questions about airline schedules, fares, reservations, and online travel agents.

Qantas Airways
http://www.qantas.com
e-mail: online form
Get online information for schedules, fares, holiday specials, and in-flight shopping.

KLM Royal Dutch Airlines
http://en.nederland.klm.com
e-mail: online form
Book directly online, read about flight services, or get frequent-flyer information.

Midway Airlines
http://www.midwayair.com
e-mail: comments@midwayair.com
Find flight schedules, special offers, aircraft information, and destination highlights from this East Coast carrier.

Japan Airlines
http://www.japanair.com
e-mail: online form
Read a complete reference guide for international passengers as well as Japanese tourist information.

Singapore Airlines
http://www.singaporeair.com
e-mail: online form
This giant site features schedules and a route map, company alliances, products and services, and other travel resources.

AirTran Airlines
http://www.airtran.com
e-mail: none available
Schedules, reservations, specials, and e-fares begin your AirTran online adventure.

American Trans Air
http://www.ata.com
e-mail: webmaster@ata.com
Lots of vacation packages, special fares, and general flight information are ripe for the clicking.

Pan Am World Airways
http://www.panam.org
e-mail: webadmin@panam.org
Historical perspective on Pan Am's aircraft, employees, and company.

Airjet Airline World News
http://www.airlinebiz.com
e-mail: airjet@airlinebiz.com
Airline news, links, and gallery with free e-mail subscription service.

AirlineRumor.com
http://www.airlinerumor.com
e-mail: airlinerumor@aol.com
Airline rumors, chat rooms, and forums to speak your mind.

ATI–Air Transport Intelligence
http://www.rati.com
e-mail: online form
Subscription-based industry resource for air transport news, airlines, and aircraft trading market information.

Aloha Airlines
http://www.alohaairlines.org
e-mail: online form
Extensive online information source for Aloha Airlines, an all-jet Boeing 737 operation that provides service to Oahu, Maui, Kauai, and the "Big Island" of Hawaii.

Aviation Directories

Pilot Shack

http://www.pilotshack.com

e-mail: info@pilotshack.com

RATING
✈✈

BRIEFING:

Keep an open mind as you click into Pilot Shack and be rewarded with quantity and quality of information.

Some of the listings are a bit hard to read. Pages take awhile to load. Site aesthetics fall into the rough category. So why the award-winning mention here? Pilot Shack throws in some good content quantity and stays true to the updating. It's a recipe for a bookmark in my book.

Sure there's some downside here. But the upside's worth your time. The laundry list of left-margin topics dips into just about anything aviation. Just a partial list steers you into airline addresses, airport information, employment information, resumé information, aviator supplies, flight physicals (listings), a calendar of events (respectably updated as of review time), travel and tour information, reading and references, gifts and collectibles, flight instruction listing, and, well, you get the idea.

Navigating through Pilot Shack, you'll find text links and your browser's "Back" button to be your new best friends. Moving around is relatively straightforward, however. Most information-related topics are broken down by specific state (yes, you'll have a host of state text links from which to choose). Anyway, when you get to your final information destination, you'll thank me—honest.

Free or Fee: Free.

Flightcraft

http://flightcraft.com

e-mail: info@flightcraft.com

RATING

✈✈✈

BRIEFING:

Leading provider of aircraft services serves up some helpful online insight too.

Certainly a leading provider of private and business aircraft services in the Pacific Northwest, Flightcraft should be spotlighted for a different reason: its informative Web site.

Yes, if you happen to be in the Pacific Northwest and you happen to need maintenance, charter, aircraft management, and the like, maybe you'll browse Flightcraft's company information offerings. However, I encourage all aviation enthusiasts to clear a space for Flightcraft.com on their favorites list solely based on the site's nice directory of aviation-only Web sites and links to emergency airworthiness directives (ADs). The aviation link list is simple, but the sites listed are bookmarkable themselves. Under the Flight Planning area, tap into sites for filing a flight plan, weather, FARs, and more. Specifically, you'll find stuff like GTE DUATS, the Great Circle Flight Path Display, the Worldwide Airport Path Finder, and DTC DUAT.

If you're more interested in the maintenance side of things, try looking under the cowling in the Search AD Notes section. Scan the Emergency ADs and locate your aircraft. Click the link that takes you directly to the FAA's listing. It's convenient and easy.

Free or Fee: Free.

PilotPointer.com

http://www.pilotpointer.com

e-mail: feedback@pilotpointer.com

RATING

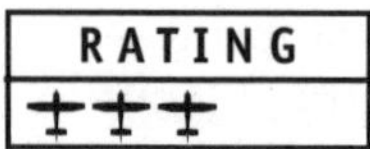

BRIEFING:

The name says it all—for online aviators, that is.

Formerly born into the Web world as the Aviator's Reference Guide (an award-winning site mentioned in the last edition of this book), the new and improved PilotPointer.com certainly has evolved into a full-blown aviation-only directory.

Building on its "aim to be your guide to quality aviation-related information on the Internet," PilotPointer.com introduces you to aviation's better sites—on one page. Where most directories spin you into a clicking frenzy, you need only scan the introductory page of PilotPointer.com's categorized links and click once. Yes once. That's it. Of course, the downside is a lack of site descriptions before you click. But PilotPointer.com is so search-friendly, it probably would be faster to click into an unwanted site and return with your browser's "Back" button than use another directory.

All your favorite categories are here, containing lots of your favorite or soon-to-be-discovered sites: aviation news, aviation search (directories), pilot jobs online, aviation weather, aviation business, flight planning, airport information, aircraft information, FBO information, aviation events, aircraft manufacturers, fuel information, aviation media, aircraft designators, airport codes, airline information, pilot organizations, and more. Yes, it's a long list. However, here's the best part: You can make it longer. Include your own favorite Web sites on the front page to easily access your sites of interest. Now that's customization.

Free or Fee: Free.

Aviation Directory

http://www.aviationdir.com

e-mail: webmaster@aviationdir.com

Briefing:

Hoping to be your "Yellow Pages" for the aviation industry—dial them up.

Self-described in its frames-driven "plain Jane" shell, Aviation Directory aspires to be your "Yellow Pages for the aviation industry and pilots around the world." Not sure where the company will end up in the future standings on the aviation directory leader board, but the company certainly has got a good thing going.

Solid, if understated design speaks to those of us who are more interested in flying, not browsing. A little less form and lots of function, please. If glitz isn't your game, Aviation Directory's for you. Left, top, and main frames handle the huge assortment of links and corresponding information well. Particularly handy are the expansion topics that appear when a broad category is selected (of course, all your favorite aviation-only topics are represented). Menus appear instantly and are easy to use. Once you've found your aviation site of choice from the Directory, it'll load directly into the main frame. Some browsing aviators enjoy this; some don't. I fall into the latter category.

Updates are reported down to the minute and seem to be fairly common. With fresh content and a minimum of dead links, any directory's bound to do fairly well. Added bonuses come in the form of current airline industry news and the ability to easily add a link to your aviation Web site for free.

Free or Fee: Free.

I Love To Fly

http://www.iluv2fly.com

e-mail: controller@pilottoys.com

BRIEFING:

Good, solid aviation directory shines with substance and speed, not flashiness.

Who doesn't? I mean, how can you not love to fly? And how can you not fall in love with an aviation directory that is so clean, fast, and complete? Answer: You can't. With kind of a magnetic appeal, I Love to Fly's directory serves up links fast and fresh, just the way you like 'em.

In my humble opinion, there just can't be too many aviation directories that get you to the good stuff in one click. Sure, there's little in the way of summaries and propaganda, but time-conscious surfers will appreciate it. What's better, all the link categories are nicely presented on a single page *sans* the heavy scrolling. Be ready for brevity, though. The main-menu options aren't many: Free Mail (sign up online for free e-mail), AV-CHAT (aviation chat room), The Hangar (classifieds), and Pilot Toys (link back to the I Love To Fly's parent Web site, Pilot Toys for aviation supplies).

So often missing the point, other megadirectories confuse and disorient the casual aviation surfer. But not here. Jump into a careful selection of sites under these categories: pilot jobs online, aviation news, aviation search, aviation weather, flight planning, aviation business search, aircraft information, airport information, aviation events, aircraft manufacturers, FBO information, aviation media, fuel information, and much more.

Free or Fee: Free.

Highabove.com

http://www.highabove.com

e-mail: admin@highabove.com

RATING

✈✈

BRIEFING:

Australian-based aviation-everything site serves up a nice little aviation directory.

Hopefully, confusion won't obstruct your VFR conditions as you fly into Highabove.com's airspace. It's much more than an aviation directory. Lots of tempting aviation goodies pull you in many directions. Some are worthy; some aren't. Feel free to frolic around when you have time. However, if it's aviation directories that interest you, Highabove.com will provide you with a satisfying arrival.

The aircrew community certainly will appreciate this site's worldwide appeal (based in Australia) and its dedication to quality links. Just look at the limited selection of four categories: major airlines, major airports, aviation associations, and aviation magazines. Oh, I suppose you could add in the "cool sites," which are pretty cool picks, but you'll find the majority of needed links in these four categories.

As I mentioned, other site clickables grace the pages too. When ground time permits, try tapping into the Cartoon Corner, Currency Converter, World Time Zones, Aviation Careers, Opinion Poll, Flight Schedules, and more.

Free or Fee: Free.

Ultimate Aviation Links

http://www.ultimateaviationlinks.com
e-mail: info@ultimateaviationlinks.com

BRIEFING:

Aviation subject searching doesn't get much more complete than this.

Tired of wading through unproductive online searches that cut into flying time? Well, roll your clicker into this breath of fresh air for to-the-point information. The folks at Ultimate Aviation Links invite you to scan their "most comprehensive listing of quality aviation-related links on the Web." Heard the claim before? Me too. Not sure if this company has the most, but I do know that its list is long and easy to tap.

The efficiently designed index page uses an "old school" left-margin button menu that takes you directly to your category of choice. And yes, there are many waiting in the wings. Reach into the growing grab bag of Web sites in message boards (four categories as of review time: aircraft for sale, flight training, aviation jobs, and general aviation), pilot shop, A&P schools, aircraft for sale, aircraft manufacturers, parts distributors, aircraft rentals, airparks, air shows/teams/races, aviation publications, FBOs, flight schools, flying clubs, aviation jobs, and more.

Just as it appears, you'll find a lot to view. Kick your feet up and explore here—there's something for everyone.

Free or Fee: Free.

JetLinks.net

http://www.jetlinks.net
e-mail: info@jetlinks.net

Briefing:
Your one-stop shop for aviation links.

Yes, it's a little bit all over the map. Smooth out your sectional, and plot your course through JetLinks.net. Will it be the aviation news or the weather? Or maybe you'd like to say a few words in the discussion forums. Whatever your reason for going wheels up with JetLinks.net, be sure to research the company's aviation links in the Resources section. You'll uncover well-represented topics, such as aeromedical reference, federal government, associations, aviation fuel, reference, schedulers/dispatch, fractional ownership, avionics, jobs, world government, maintenance support, communication services, and more.

A subtle yet nice twist is "star" ratings next to your aviation links in certain categories. Yep, JetLinks.net also gives you a heads-up before you click and offers its recommendation in the form of "stars." Whether you agree or not with the rating, it still gives you a mild benchmark in deciding to click further or throw on the air brakes.

Again, more than just an aviation link farm, JetLinks.net dabbles in other things too. Business aviation news, weather, flight planning, discussion forums, chat, and more grace these online skies.

Free or Fee: Free.

AviationPage.net

http://www.aviationpage.net

e-mail: online form

RATING

BRIEFING:

Aviation links, companies, and information tied up in a perfect online package.

Handsome design first attracted me to AviationPage.net. And once I got clicking around, I quickly added a bookmark. This directory's got it all. High-powered Web layout skills *and* quality content. Now that's a knockout combo.

Something's sure to strike your fancy here. Pleasant navigational menus draw you into a huge Web Directory chock full of categorized listings. From Aircraft Acquisition to Maintenance Repair and Overhaul, it's here. Plenty of clicking. Lots of links. Find your category of choice, and often you'll be invited to narrow it down with subcategories. Actual site links come complete with title, summary, and rating based on viewer response.

But wait, there's more. Much more. Jump into an FBO Directory to, you guessed it, search through a list of fixed-based operations worldwide. Compute-a-Plane is the place to buy or sell anything aviation-related. Aviation World Directory is your one-stop shop for aviation companies. Aviation Consultants steers you into a collection of links to services in every area of aviation. And Jet Lease puts you in touch with commercial aircraft and engines for lease and sale.

Free or Fee: Free, even to add listings.

Aeroseek

http://www.aeroseek.com

e-mail: contact@aeroseek.com

Briefing:

Good, solid aviation directory shines with substance, not style.

Yes, there are actually many aviation-only directories—all pretty good; some really good. But quantity, quality, and variety are the differentiating factors. Put aside this directory's garden-variety design and open your mind to its labyrinth of categories. Nice, thoughtfully specific topics guide your direct course to air carriers, aircraft acquisition, aircraft operation, aircraft specific (sites dedicated to specific aircraft), classifieds, databases, general aviation and sport aviation, images, publications, safety, and more.

Okay. I can almost hear you asking, "Yeah, but what about quantity in each category?" And the answer is: sufficient but not overwhelming. At review time, about 3700 site links graced Aeroseek's pages.

Top 2 percent sites (loosely described as "cool"), a nifty keyword search, and more round out your page options.

Free or Fee: Free.

FlightSearch

http://www.flightsearch.com

e-mail: comments@flightsearch.com

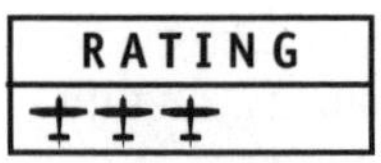

BRIEFING:

Shining like a rotating beacon in fog, FlightSearch gives worthy pointers to aviation sites. But you'll be impressed with the extras.

With FlightSearch, your aviation Web location needs are met. However, its flair for extras keeps you coming back.

A nice, easy-to-use search box steers you toward the store of linked offerings. From adventures to airport restaurants to services, it's all here. Aircraft, airlines, getaways and resorts, and more contain a respectable variety of sites, but reach over to the right-hand menu and get a bit more for your browsing time. Find your favorite airline links at a glance under Major Airlines. Then introduce your kids to a friendly Web site created specifically for them called PlaneKids. Or skip to FlightLink, which directs you to an interesting enterprise of aviation-related free banner ad listing.

Free or Fee: Free.

FlyByWeb

http://www.flybyweb.com

e-mail: none

RATING

BRIEFING:

A good aviation source for sites. Take a peek.

Tucked away near Davenport, Indiana, a couple of guys (one's a pilot, one's a Web guru) are spinning a pretty resourceful aviation web. FlyByWeb is the resulting product from "countless hours of research for the benefit of aviation enthusiasts everywhere." There's no question that this compilation of categorized links did take some time. It also goes without saying that it certainly is a benefit to browsing aviators.

The simple yet mildly appealing design is anchored by a heavily filled left-margin frame. Your menu options are many, so settle in with nourishment at the ready—you may be here awhile. Get to your aviation destination quickly with categories that are alphabetized and organized into subtopics for fast searching.

Although each category lacks huge volumes of listings, you will uncover variety and quality. Use the left menu bar to begin displaying your selections in the target (right) window. Find the good stuff in airports, airlines, accident reports, aircraft sales, associations, employment, flight training, fun links, parts, professional services, software, and more.

Free or Fee: Free.

Rising Up Aviation Resources

http://www.risingup.com

e-mail: online form

RATING

BRIEFING:

Yes, the directory's good, but the extras keep you coming back.

Like a hot mountain thermal, Rising Up Aviation Resources lifts the link list directory into a new dimension. Sure, the aviation links are here, carefully categorized and subcategorized into a list of clickable summary descriptions. Flawless by directory standards, it's fast, accurate, informative before you click—and maneuverable. However, the extras push Rising Up into the elite hangar of award winners. Why? Because it employs that award-winning team of ease of use, speed, and clean layout. Add in some extras that aren't always found at directory sites, and you've got yourself a bookmark addition. Quick links to aviation books, medical examiners, and weather take their place on the startup screen, as do discussion forums, airplane performance specs, FARs online, and practice FAA tests.

Rising Up's main reason for success—comprehensive directory information—earns accolades for its many levels of subtopics and essential summary descriptions of each site—a must for the time-conscious browser. Current main topics that are further broken down include airports and airlines, clubs and associations, aviation careers, government, weather, pilot resources, sport aviation, and more.

Free or Fee: Free.

4Aircraft.com

http://www.4aircraft.com
e-mail: online form

RATING

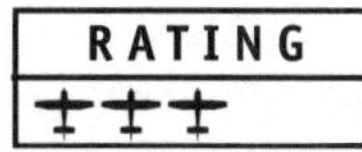

BRIEFING:

4Aircraft.com joins a host of other topic-specific Web guides from the people at 4Anything.com.

A specialty Web guide from the 4Anything.com Network, 4Aircraft.com is exactly what its name implies—a jumping off point to aircraft-related sites. While its supersonic loading speed will surprise you, its related aviation sister sites will delight you. When you're through scanning the aircraft-specific stuff, stroll over to the equally enriching sites of 4Airinfo.com, 4Airline.com, 4Charter.com, 4Destinations.com, 4Pilots.com, and 4Travel2.com.

With sparkling clean page presentation, your aircraft-related options become crystal clear. For example, links you may find under Aero-Marts include Aerotrader, Aircraft Shopper Online, AAA Airplane Exchange, Boeing, and more. Under Publications and Associations, expect to find stuff like National Business Aviation Association, Scramble Magazine, Aircraft Owners & Pilots Association, Sport Aviation, and more. Information and Services promotes sites such as Garret Aviation, NOAA's Aircraft Operations Center, Duncan Aviation, and more. You get the idea. Lots of links can be found for aircraft-inquiring minds.

Free or Fee: Free.

Aerospace Mall

http://www.aerospacemall.com

e-mail: online form

BRIEFING:

Aerospace Mall opens its cyber doors to linked resources and information.

Different from a real-life megamall, Aerospace Mall makes shopping for aviation-related information easy—and you don't even need to find a parking spot. Although buying and selling are not the focus of this mall, information exchange is. The mall directory gets you to your topic of choice quickly with no crowds or waiting.

Although some superfluous stuff floats around in the mall, of real interest are the industry news and the mall directory. Sure, if you're looking at significant ground time, you can browse through the corporate profile, job board, discussion board, games, and more. However, the good stuff, the meat of the mall, is information.

Stop in the directory for pointers to aircraft manufacturers; space; avionics; air transportation and terminals; aviation, airport, and business services; FBOs; fuel systems; reference; and more.

Free or Fee: Free.

Smilin' Jack

http://www.smilinjack.com
e-mail: jack@smilinjack.com

RATING
✈✈✈

BRIEFING:

A quality aviation/airline directory that keeps you "smilin'."

Thoughtfully void of giant pictures and silly graphics, this sleek aviation directory creates minimal drag and maximum lift. As noted on the page itself, it was designed for easy loading and navigation. A simple menu guides you into efficiently organized content for airlines (a huge list of airline links), fun flying (hyperlinks to popular sites), servers (more links, mostly directory sites), airports (comprehensive list of major international stuff), shopping, home pages (miscellaneous smattering of favorites), and weather (over 40 sites make you weather wise).

By the way, in case your inquisitive mind really wants to delve into the background of Smilin' Jack, here's the scoop: He was a character in a newspaper comic created by Zack Mosley that ran from 1933 through 1973.

Free or Fee: Free.

Captain Bob's Pro Pilot Page

http://www.propilot.com

e-mail: postmaster@propilot.com

BRIEFING:

A simple, easy-going collection of aviation stuff for your viewing and listening pleasure.

Captain Bob's Pro Pilot Page isn't about giant, slow-loading plane pictures. Nor is it particularly fancy with modem-stalling graphics. It is, however, a solid, simple resource for pilots and aviation enthusiasts. Period.

Captain Bob (a California captain for Skywest Airlines), with his cyber enthusiasm, has amassed a nice collection of aviation-related links and resources. The best parts? The page doesn't take forever to load, it's free of memberships, and all the hyperlinks seem to work without fail.

When you can catch your breath from all those giant, time-wasting aviation sites, check in here for hordes of aviation links, airline/airport-specific sites, weather resources, a pilot bulletin board that's full of entries, an interesting commentary on regional airline careers, background on Captain Bob, lots of live video cams of Doc's FAR Forum, and more.

My personal favorite: the live links to DFW Air Traffic Control and Chicago Air Traffic Control. To tune in, though, you will need a RealAudio player or equivalent. Most newer browser versions have one built right in.

Free or Fee: Free.

DeltaWeb Airshow Guide

http://www.deltaweb.co.uk/asgcal

e-mail: webmaster@deltaweb.demon.co.uk

RATING
✈✈

BRIEFING:

Conscientiously updated world-wide guide to airshows.

As noted in the guide's text descriptions, airshows, by their very nature, are apt to change—sometimes radically—at the last minute. DeltaWeb's Airshow Guide does a great job combining current updates with well-oiled page organization.

The calendar is arranged by region (as of review time, regions include the United Kingdom, North America, Europe, and the rest of the world) and further subdivided into months. Simply find your region and pick a month. Airshows will be listed in date order with location, short descriptions, and contact information. For example, a June airshow listing might read: "Saturday 7, Scott AFB, Illinois. HQ Air Mobility Command Base Open House and Airshow USAF 50th and Salute to the Berlin Airlift, Thunderbirds and Golden Nights. Tel: USA 618-256-1663."

Navigation throughout is painless. From each page, additional months for the same region are readily available via your clicking fingertips. Conversely, other regions for a chosen month are standing by simply with a click of the regional icon.

Free or Fee: Free.

International Aviation Directory

http://www.infomart.net/av

e-mail: acadir@infomart.net

RATING
✈✈

BRIEFING:

Complete international aviation directory that lists, searches, and categorizes without charging a penny.

Bursting at its cyber-seams, the American & Canadian Aviation Directory is a one-stop resource with over 54,000 aviation companies listed. Conveniently available via a custom site-searching tool, this aviation-only smorgasbord launches you into online aviation with only a couple of clicks.

Once you've maneuvered around the site's miscellaneous banners, your search form should begin to come into view. Simply enter any or all pertinent query data—from category to telephone number—and search. The directory does the rest. After it chunks through a fairly complete database, on the other end of your modem, up pops your results. Category curious? Here are the highlights (with a lot of other stuff in between): aircraft charter/nonscheduled air transportation; aircraft dealers and brokers—retail; airports, flying fields, and airport terminal services; flight training, airports and aircraft maintenance; aircraft finance; aircraft storage; and more.

Itching to get your own aviation site noticed? Add your own Web site address and company details here—it's free. Hey, when did you last read the words *free* and *aviation* in the same sentence? That's what I thought. Take advantage!

Free or Fee: Free.

AirNemo

http://www.geocities.com/CapeCanaveral/4285
e-mail: airnemo@geocities.com

BRIEFING:

An airline-information diamond in the rough that's hard to find using traditional search engines. You'll be glad I did the digging.

It's never late. Departures are always on time—24 hours a day. You won't even need to sprint to the gate. Just grab a mouse and settle into a first-class seat. Destination? AirNemo—The Best Links to Airline Sites. Dedicated to the air transportation industry, AirNemo boasts a knowledgeable site author behind the scenes who works for an international airline in Brussels, Belgium. As such, you might have guessed French descriptions are conveniently standing by if English won't do.

Stylishly unobtrusive graphics and icons dominate the visual horizon, while excellent organization makes for frustration-free navigation. Look for convenient site codes like "new" flags, "official/unofficial" designations, and more. And tying the pages together with invisible packing tape is a nice omnipresent, bottom-page frame menu.

At first glance, you may think that airline links and stats make up the entire site contents. While it's true, you'll find commercial aircraft characteristics about the B747-400, A340-200, MD-95, and others, I urge you to dig deeper. The corporate aircraft characteristics, acronyms, conversion tables, links, sites, and codes are fascinating. Even if you disregard all the preceding recommendations, do tap into the weekly Air Bulletin. It's simply a well-done online gem for aviation, air travel, and related issues.

Free or Fee: Free.

Space.com

http://www.space.com

e-mail: thoughts@space.com

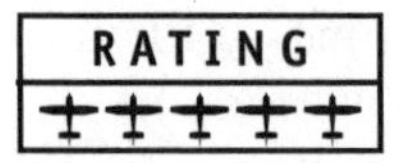

BRIEFING:

A masterful mix of multimedia gizmos and futuristic design for space-age enthusiasts.

Almost as infinite as its subject, the Space.com site touches down on an unending collection of goodies. News, events, history, and education certainly create a solid core. However, the true site mission revolves around a multimedia mix of unearthly content and out-of-this-world presentation. Contained within an electrifying display of design, fun gadgetry includes NASA audio/video and coverage of space-related news, events, missions, and more. Space.com's multimedia archive is a self-described one-stop shop of videos, audios, and FLASH animations culled from Space.com's own library and from NASA archives.

Once you're up on the most recent reports and rendezvous, go back a few light years into a complete look at space history. With stunning visual images and rare NASA footage, Space.com captures the excitement of exploration. Get space privy with astronaut biographies, a space history timeline, mission footage and photographs, personal accounts, and planetary probes.

This heavenly compilation is enough to make you weightless with space-driven ecstasy. If the page design doesn't launch you, the content will.

Free or Fee: Free.

Russian Aviation Page

http://aeroweb.lucia.it/rap/RAP.html

e-mail: agretch@aeroweb.lucia.it

RATING

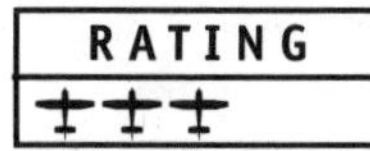

BRIEFING:

Russian aviation riches aplenty. Enjoy your flight.

Getting globally gregarious, I've accidentally stumbled across one of the most exhaustive informational sites on the Web today. The Russian Aviation Page is utterly bottomless in depth and pleasantly surprising in breadth.

Faster than a MiG-21 flyby, a flurry of well-designed aviation information bursts onto your screen from the moment you type the address. Even if Russian aviation stats and features aren't high on your bookmark list, I nevertheless encourage a visit. There's news, Russian aviation FAQs, museum review, digital movies, an image archive, monthly Russian aviation trivia, and countless links.

Although page contents are carefully arranged and presented with award-winning style, there's almost too much to digest in one modem visit. Plan accordingly. Multiple menus, helpful search engines, and "new/update" tags offer the best assistance as you maneuver through this Russian wonder.

Skimming the Russian surface, some must-see features include chronology (development of the aviation industry in Russia and the Soviet Union, 1916–1946); the Russian version of *Top Gun*; the Russian Superfortress—the Tu-4 Story; and many downloadable movies.

Intrigued by Russian sites? Researching Russian aviation? Stow away on a Tu-142, and join other enthusiasts here.

Free or Fee: Free.

AERO.COM–Future of Aviation

http://www.aero.com

e-mail: info@aero.com

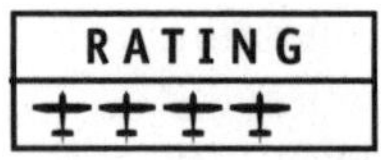

BRIEFING:

A worldwide reference resource of magazines, newspapers, and newsletters.

This nice little directory of links features a couple of extras you wouldn't normally find: ballooning, parachutes, and a "helicoptorial," for instance. Step into On-line Reading for an eye-opening list of magazines, newspapers, and newsletters. The AERO.COM Departments Directory features a fairly comprehensive topic list of art/photography, an experts page, a research center, flight planning, a flying shop, museums, news, organizations, and more.

You also can tap into the FAA Online for *NorCAL, SoCAL, Hi Desert Airman, Pacific Island Flyer,* and *The Aviation Yellow Pages.*

Free or Fee: Free. Get on the mailing list for updates.

The Air Affair

http://www.airaffair.com

e-mail: wingnut@airaffair.com

RATING

✈✈

BRIEFING:

An online directory that gives you stunning efficiency by people who know what they're doing.

It's always refreshing to find an aviation site that truly has your interests in mind. The Air Affair does have the news, photos, links, and event listings that most "e-zines" have. However, once you begin maneuvering through the selections, you notice the distinct difference of complete organization and speed. Put another way, VORs work fine most of the time, but it's nice to just punch the GPS a few times and get to your destination with ultimate ease. The Air Affair is designed to be browser-independent (nice) and features a lack of frivolous graphics (not necessarily a beautiful piece of Web art—but it's fast), thoughtful searching engines for fuel prices, a local flight-training locator, and a well-thought-out page of navigational aids.

Bookmark this site for the ultimate in events, aviation fuel prices, an aviation library, flying destinations, a flight-training locator, other aviation Web sites, Ask Propellerhead, and a photo gallery.

Free or Fee: Free.

Landings

http://www.landings.com

e-mail: landings@landings.com

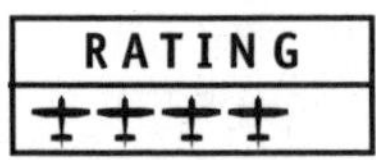

BRIEFING:

Award-winning directory/ database for every aviation subject imaginable.

Dubbed "aviation's busiest cyber-hub," Landings guides you into a huge, expertly maintained collection of information goodies. Sophisticated search engines move you through FAA and Canadian regulations, AIM, the pilot/controller glossary, service difficulty reports, air-worthiness alerts, NTSB briefs, N-numbers, the FAA Airman Database, airman knowledge test information, and more. There are good links all over the place. Directory subjects include aerobatics/flying, aircraft sales, aircraft service/parts, manufacturers, airlines, air-ports, aviation BBSs, aviation images, aviation news groups, avionics, companies, flight schools/FBOs, flight planning, general aviation, GPS/technologies, hang-gliding/paragliding, helicopters/gyrocopters, home-building, military, museums/history, publications, soaring, travel tours, and worldwide weather links that would leave goose bumps on a meteorologist.

The best part of this enormous informational grab bag is at-a-glance navigation. You'll find that the concise "front page" offers news and quick-click menu cate-gories. Check in here often. The company maintains and updates regularly.

Free or Fee: Free.

AirNav

http://www.airnav.com
e-mail: pas@airnav.com

BRIEFING:

A virtual smorgasbord of free airport, navigational fix, and fuel data.

Okay. Enough flying frivolity. Let's get down to some serious navigational aids. AirNav pulls critical flight-planning data out of cumbersome FAA publications and displays it on a virtual silver platter. Free and at your fingertips, you'll breeze through current facts, figures, and frequencies. Click on Airport information for an amazing array of searchable airport data (way more than what you would find in the Airport and Facilities Directory). Navaid Info gets into the nitty-gritty of radio navigation (VORs, NDBs, TACAN, marker beacons, etc.). Fix Info provides enroute fixes, airway intersections, and waypoints ("From AAAMY to ZZAPP"). Then figure in your pit stops with the Fuel Stop Planner. It's a well-organized site to begin your well-organized flight planning.

Note: It's important to remember that the information contained here is *not* valid for navigation or for use in flight. It is simply provided as a tool. Use data at your own risk.

Free or Fee: Free.

Cyberflight

http://www.cyberflight.com
e-mail: ooblick@cyberflight.com

BRIEFING:

A Maryland-based East Coast directory site that's in complete disarray. The redeeming quality? It's roll-out-of-your-chair funny.

Where to start describing Cyberflight? Um, it's an autobiographical diary of perilous cross-country expeditions and a hodgepodge of links (some even non-aviation-related). Toss into the cyber-blender a little sarcasm mixed with rambling commentary, and you've got an amazingly entertaining online place to visit. Don't forget to check out the embedded photos—they're a riot as well.

Once you cut through numerous solicitations for "lots of money" (donations, I guess), you'll learn where to fly to find food, to find "cool" airports, to find what's new at Cyberflight, to learn to fly, to shop, to learn about aviation weather, and to stumble across a host of miscellaneous stuff in Fun Stuff About Us.

Although there's actually some good information here, it's a great place to visit for some old-fashioned hysterical high jinks.

Free or Fee: Free.

AeroLink

http://www.aerolink.com
e-mail: info@aerolink.com

BRIEFING:

An award-winning global information source that overflows with clickable mastery.

Fashionably designed and desperately needed, AeroLink really shines with its well-oiled search-engine machinery. Peer into its linked offerings, and you'll get goose bumps over its simple yet practical efficiency.

Three menu options provide clickable doorways into this aviation "link farm": Who/What/Why, Add Your Link, and Feedback. (First-timers may want to check into Who/What/Why before jumping into the list of hyperlinks.) Major search categories in which you'll find yourself pleasantly inundated include airlines and airports, government, industry, industry OEM, piloting, education, and safety.

Next, step into the specific subcategory. Yes, the refined searching is endlessly exhaustive. Finally, for adventurous surfers who happen to have their own aviation-related home page, Add Your Link does just what it says—easily.

I know. You're thinking no site can be this good. Well, just type in the address and thank me later.

Free or Fee: Free.

Aero Web: Aviation Enthusiast's Corner

http://aeroweb.brooklyn.cuny.edu

e-mail: air-info@brooklyn.cuny.edu

Award-winning aviation-related features. Stop by for content—it's worth the ride.

Meander through the contents and you'll uncover this site's reason for being: variety. Although it's pleasant graphically, you'll really appreciate the nice smattering of useful information. Maintained by the hardworking volunteers at Brooklyn College, categories include a museum index by location, an aircraft locator by type and manufacturer (here you'll find a long list of aircraft with some combination of descriptions, performance, specs, and museum display location), an airshow location and performance index, airshows by month, aviation history features, aviation records, some local New York stuff, and more.

Even while skipping among Aero Web's offerings, you'll never lose your place—thanks to handy page menu headers. So let loose and have some fun here. True enthusiasts will learn something.

Free or Fee: Free.

Women in Aviation Resource Center

http://www.women-in-aviation.com

e-mail: online form

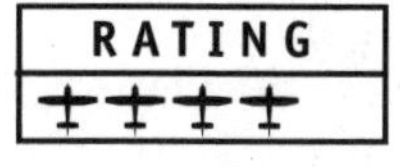

BRIEFING:

This female-only forum opens the door into a world of women-in-aviation information.

With similar veracity as Amelia Earhart, the Women in Aviation site sheds important light on resources that might have appeared dim. This site firmly establishes an educational link to seemingly hard-to-find resources such as books, education and training, museums, networking online, organizations, publications, upcoming events, women in business, and more.

Cleanly arranged and quick to access, this cyber-forum makes networking convenient. There's a list of women-in-aviation contacts, complete with titles and e-mail accounts. Or participate in open discussions with the site's online forum. Whether you're here to browse or choose interactive involvement, straightforward directions and descriptions abound.

Free or Fee: Free.

GlobalAir.com

http://www.globalair.com

e-mail: webmaster@globalair.com

BRIEFING:

Good internal database makes searching for aviation sites and classifieds pleasantly efficient.

Marvelously redesigned into a dominating directory, GlobalAir.com speeds you to the Web's best in aviation with style. Useful topics and subtopics narrow your focus in the aviation Internet directory for FBOs, airports, brokers/dealers, weather, insurance, FAA stuff, financial services, parts, flight schools, pilot services, reservations, and more.

Specifically, plane and parts enthusiasts will find a satisfying collection of classified ads—easily searchable with a text box prompt, QuickFind categories, or a laundry list of infinite sections. Good thing the site makes searching easy because loads of listings are the rule here. From batteries to floats and consultants to warbirds, the variety is pleasantly satisfying.

Events calendar, e-store, and more give you even more reasons to add a bookmark.

Free or Fee: Free. Fee for listing an ad.

Bookmarkable Listings

Alex's Helicopter Home Page
http://www.geocities.com/CapeCanaveral/3838
e-mail: amartins@abordo.com.br
Model and full-size helicopter information and resources.

Air Cargo Online
http://www.cargo-online.com
e-mail: webmaster@cargo-online.com
Central cargo database of available air charter capacity worldwide.

Calin's Aviation Index
http://www.calinsai.com
e-mail: mail@calinsai.com
A central source of aviation contacts and news.

Airshow.com
http://www.airshow.com
e-mail: none provided
Collection of current airshow schedules, performers, and related information.

Airship and Blimp Resources
http://www.hotairship.com
e-mail: webmaster@myairship.com
A volunteer effort to provide airship information to newcomers and veteran aeronauts.

Wings of History
http://www.wingsofhistory.org
e-mail: webmaster@wingsofhistory.org
Link resource for antique aircraft, including museums, organizations, and directory sites.

Aviation Business Center
http://www.airsport.com
e-mail: webmaster@airsport.com
Links to businesses supplying products, services, and information to the aviation community.

NOTAM
http://www.notam.com
e-mail: webmaster@notam.com
Diversified directory for airports, airlines, classifieds ads, general aviation, and more.

The Aviation Home Page
http://www.avhome.com
e-mail: see site
Completely searchable aviation directory easily speeds you to your destination.

HeliCentral
http://www.helicentral.com
e-mail: admin@helicentral.com
Fantastic all-helicopter link directory.

Globemaster US Military Aviation
http://www.globemaster.de
e-mail: webmaster@globemaster.de
Extensive U.S. military aviation database containing flying units, air bases, and tail codes.

Airportclassified.com
http://www.airportclassified.com
e-mail: webmaster@airportclassified.com
Huge collection of aviation classified ads searchable by category, location, date posted, and keywords.

Aviation Organizations and Associations

Airline History Museum at Kansas City

http://www.saveaconnie.org

e-mail: info@airlinehistorymuseum.com

Briefing:

Fantastic Web-style documentary of a Lockheed Super G Constellation (Connie) and other bygone aircraft.

I'll readily admit that I'm not all that familiar with the Lockheed Super G Constellation. However, as I clicked into this "Save a Connie" site, I almost couldn't help but be intrigued. In 1986, a group of dedicated Kansas City–based aviation enthusiasts decided they were going to find, acquire, make flyable, and restore to "like new" condition a Lockheed Super G Constellation. The short version of their story is that they achieved their goal, and it's available for viewing online and in person. The longer story is written well and presented perfectly with pictures.

The Connie, however, is not the end of this Web tale. Two other restored aircraft represent themselves proudly here too. Click in to meet a beautiful DC-3 and a Martin 404 (one of only three flying in North America). As you would expect, stats, pictures, and history are prevalent throughout. So set aside some quality reminiscing time.

Although more of a perk than a necessity, multimedia presentations of this site's aircraft make viewing fun. Get 360-degree interior views, movie clips, and audio WAV files of the Constellation. You can even download a realistic flight model of the Connie for use in Microsoft Flight Simulator!

Free or Fee: Free.

U.S. Navy Blue Angels Alumni Association

http://www.blueangels.org

e-mail: baalumniassn@aol.com

RATING
✈✈

BRIEFING:

Pictures, history, and perspective from one of the world's most popular air show participants.

Even with lackluster site design and looping military tunes in the background, The U.S. Navy Blue Angels Alumni Association proves that content is almost always king.

Once you embark on your Blue Angels journey here, takeoff might feel a little unstable. However, climb out of ground effect and start clicking through the contents (get to the site map for best searching results): Blue Angels Aircraft (great 53-year evolution of the Blue Angels aircraft), Blue Angels Crew Pictures, Blue Angels Fat Albert Marine Crews, Blue Angels History, Blue Angels Aircraft Maintenance, Our Photo Album, and more. Yes, the site's mostly pictures and limited descriptions. In the Blue Angels History section, however, you'll find some excellent stories and background written by various Blue Angels alumni.

Design and organization of the U.S. Navy Blue Angels Alumni Association Web presence leave a fair amount for improvement. However, this obviously collaborative, seemingly volunteer-driven effort still gets quickly to the good stuff with big yellow buttons and a handy site map–style table of contents. If you plan on revisiting, you'll want to first check into the What's New section for updates sorted by date. It goes without saying that if you happen to be a former team member, you'll want to review the Alumni Association information.

Free or Fee: Free.

Not a team member? That's okay. You'll enjoy the better part of a day flipping through fun pictures, history, and happy background tunes.

Wings of Freedom–The Royal Canadian Air Force

http://www.rcaf.com

e-mail: webmaster@rcaf.com

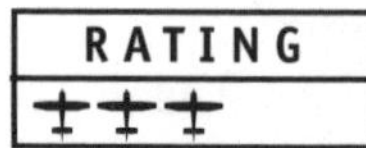

BRIEFING:

Unofficial Royal Canadian Air Force tribute is officially an award-winner.

Though it's not affiliated with, nor endorsed by, the Canadian Department of National Defense or the Air Force Association of Canada, this privately developed tribute to the Royal Canadian Air Force simply must have its day in the sun. Organization and site design are so well thought out that you just can't get lost. No need to dial in the GPS or follow the moving map. Simply view the main introductory page to get all your clickable options at a glance. Even an In the News menu of assorted links takes you to current topics of interest.

A quick-jump menu, left-margin rollover menu, and main page features guide you to your first destination. From there you risk losing yourself to intrigue while reading and learning about this fascinating organization. Delve into the history and begin studying The Formation, The War Years, The Cold War, The Unification, and The Present. Sure, there's quite a bit of reading. However, gallery photos abound too, especially within aircraft lists. Tap into this site's aircraft encyclopedia for the Arrow, Cunuck, Mustang, Sabre, Starfighter, Voodoo, Vampire, C-5, Dakota, Expeditor, Mentor, Chipmunk, Silver Star, and many more. Descriptions and presentation are entertaining, enlightening, and inspiring.

Many more site features take the form of Information (about the Web site itself), Message Board, Personnel Database (for Canadian service men and women only), Wallpaper (decorate your desktop), Clipart (marginal-quality aircraft clip art), and Links.

Free or Fee: Free.

Aviation Link Exchange

http://www.aviationlinkexchange.com
e-mail: staff@aviationlinkexchange.com

RATING

BRIEFING:

Promoting your aviation-focused Web site? Start here.

Smart Web marketers know that link and banner exchanges can be a gold mine for generating traffic to a Web site. If you're an aviation Web site owner, developer, or Webmaster, I've got just such a service for you. If you're new to the concept, here's a head's up: Link and banner exchanges are collections of individual Web sites that trade banner advertising space on related sites (in your case, aviation) for *free*. And in most cases the rate is a one-for-one banner exchange. For every one banner you show on your site, you receive a banner on another member's site.

Now, on to a few Aviation Link Exchange specifics. Site design falls into the excellent category due to its simple, easy-to-view layout. Everything functions normally, as it should. Menus are where they should be. And information is presented well. The services are also spelled out clearly, with membership-level options that are easy to use.

Aviation Link Exchange and others like it are a welcome site for those looking for a little Web-promotion ammunition. Take advantage, it's free.

Free or Fee: Free and fee-related services available. See Web site for details.

Canada's Air Force

http://www.airforce.dnd.ca/airforce

e-mail: ae737@issc.debbs.ndhq.dnd.ca

BRIEFING:

Officially captivating. Officially informative. Officially Canadian.

Yes, this one's official. The huge cyber presence for the Canadian Air Force pleasantly bombards you with information about today's Canadian Air Force, its organization, aircraft, and history. My online visit included a peek into Air Force Life, Defending Canada, Search & Rescue, Combat Capable, Facts and Stats (Aircraft), Headlines (in the News section), Timeline and Articles (in the History section), and some external links.

Perhaps the most fun, though, can be had during the virtual reality tour of the CF-18 fighter. Tour the cockpit, and "walk around the aircraft." Even an audio explanation enriches your 3-D journey. During your tour, don't forget to click on Hot Spots to bring up captioned close-up images. Next, grab your flight suit and stroll into the Video Library. "Slider" takes you on a virtually exciting ride as he rolls the aircraft, pulls G's, and performs other maneuvers.

Another favorite? The Pre-Flight Tour. This ain't your typical Cessna 182 preflight. Click in to see what it takes to get a pilot into the air with this cool interactive slideshow. Great pictures and descriptions guide you through this fascinating process.

Free or Fee: Free.

International Air Transport Association (IATA)

http://www.iata.org

e-mail: webmaster@iata.org

BRIEFING:

International airline industry organization warms up to its members with a well-crafted Web presence.

Not too sure how many aviation surfers will find the IATA site of interest. Its stellar online presence, nevertheless, warrants folks to take note. Worldwide in scope, the IATA represents and serves the airline industry. Just for the sake of argument, let's say you want to learn about IATA. This information-packed Web brochure catalogs everything you'd want to know. You literally could spend hours pouring over IATA's descriptions of mission, goals, values, history, annual reports, press releases, and speeches. Careers (with IATA and member airlines) and the Aviation Links section may appeal to a more broad audience.

I think we can agree that every major aviation organization needs a voice on the Web. And IATA certainly has captured the essence of how to use its online muscle. Thoughtful layout, professional design, and perfect navigation make your clicking experience with IATA a pleasant one. You'll notice that every page offers a main, top-left toolbar for major site topics. Help, FAQs, Site Index, and Contact Us are also standing by conveniently.

Free or Fee: Free to view organization information.

International Council of Air Shows (ICAS)

http://www.airshows.org

e-mail: icas@airshows.org

Briefing:

Air show schedules, facts, message boards, and photos are on tap at ICAS—even for nonmembers.

Founded in 1968, ICAS protects and promotes the interests of its member organizations in the growing North American air show marketplace. Its well-developed air show in cyberspace flies loops and barrel rolls over most air show–driven Web sites. Perfect layout and presentation rival that of a 300-knot flyby: an awe-inspiring site that you just don't see everyday.

Content is a show stopper, too. The ICAS Air Show site could have made everything password-protected for members only. But it didn't. Even casual-browsing nonmembers can check into air show schedules, air show facts, a photo gallery (just getting started at review time), lots of message boards, and feature articles from the printed *Air Shows Magazine.*

Most everything site-wide is free and readily clickable. Members, however, gain special access to stuff like the ICAS Member Database, FAA forms and documents required for air shows, guidelines for air show pyrotechnics and special effects, guidelines for jet vehicle demonstrations, the ICAS Air Show Manual, ICAS by-laws, and more.

Free or Fee: Free to most information. Fee for ICAS membership.

National Association of Flight Instructors (NAFI)

http://www.nafinet.org

e-mail: nafi@eaa.org

BRIEFING:

NAFI climbs into the right seat with flight instructors.

Founded in 1967, NAFI dedicates itself to "raising and maintaining the professional standing of the flight instructor in the aviation community." With an honest, thorough approach, NAFI online jumps into the cockpit with instructors and guides them through important organizational benefits, such as recognition programs (Master Instructor Designation Program and others), safety and technical information, CFI hiring tools, industry news and links, and more.

Of particular interest to a cross section of companies, instructors, and students probably would be the Post a Job and Find an Instructor sections on the site. Those companies interested in filling instructor positions need only fill in job details and send to NAFI. Prospective or current students shopping for a qualified collection of instructors should find the Find an Instructor easy to use and thorough. Simply pick your state and review the long list of instructor information, sorted by city.

Free or Fee: Free to search for instructors and post an instructor job. Fee for NAFI membership.

WASP-WWII

http://www.wasp-wwii.org

e-mail: nancy@wasp-wwii.org

RATING

✈✈✈✈

BRIEFING:

The rich history of Women Airforce Service Pilots during World War II is a fascinating read.

Over 50 years ago, exactly 1857 young women pilots came from all over the United States to become Women Airforce Service Pilots (WASPs) during World War II. WASP-WWII online brings the rich WASP history to light to honor their contributions. Yes, this well-designed Web tribute has the photos and history. However, if that's all you see, you've missed the point. A huge compilation of stories and records, audio and words, tells quite a gripping story.

Circumvent the unnerving array of banners, Web rings, and ad links on startup to begin the journey. Once inside, the clutter turns to compiled history, nicely organized. The official menu for WASP invites you to experience Arcade (Java games and fun), Records (research documents and statistics), Gallery (pictures, drawings, audio, and video), Resources (more research material), Interact (message board, mailing list, chat room), WASP Stories, and more.

Keeping the interest up for the casual Web wanderer is WASP's attention to detail (minimal typos) and great writing among the many snippets, articles, reports, and historical perspectives. Old-time photos are captivatingly enlightening.

For a real thrill, download the video for a brief history of the WASP.

Free or Fee: Free.

National Warplane Museum

http://www.warplane.org

e-mail: nwm@warplane.org

RATING
✈✈✈

BRIEFING:

The V-77 Stinson? Alive and well, thank you, via this online museum.

Physically located at Elmira-Corning Regional Airport, New York, and on virtual tour via cyberspace, the National Warplane Museum site is captivating. Graphically motivated, behind-the-scenes buffs have obviously zeroed in on exceptional design. Better still, this mix of visual treats doesn't make a mockery of your modem. Loading is fairly efficient, multiple types of menus are always at hand, and the "cockpit" aircraft searching feature is fun.

Sure, there's the usual on tap, such as Museum's Mission, Sorties (where the museum's planes have been and where they are moving around to), Enlistment Form for Members and Volunteers, News, Pilot Interviews & Stories, links to other sites, and more. However, where you're going to want to spend some time is the online plane collection. Just find your favorite in the pop-up window and enjoy. From an F-14 Tomcat to an L-3 Grasshopper, it's all here via mouse and modem.

Free or Fee: Free.

Aviation History On-Line Museum

http://www.aviation-history.com

e-mail: lpdwyer@aol.com

BRIEFING:

Tour this online museum for a cyber-stopover into the archives of aviation history.

Aviation enthusiasts young and old will revel in the online convenience of this interactive museum tour. No dusty textbooks here. Just pixel-quality pictures and lots of insightful description. Those with a fancy for flying will quickly become enamored with a nice collection of historic aircraft.

Neatly indexed by manufacturer and country, a quick click launches you into a concise aircraft summary—with an option for the full-text version. Although the punctuation-challenged descriptions run amok, the facts and stories are still fascinating. The hand-picked index of aircraft includes more than a few of my favorites: P-26 Peashooter, Mosquito, P6 Sea Master, Spitfire, P-47 Thunderbolt, B-24 Liberator, P-40 Warhawk, and The Concorde. Yes, even the Corcorde. Hey, it's still a history-making phenomenon.

Once you've saturated your curious mind with aero wonders, there is still more to study. Mouse over to the left-margin buttons for Engines, Early Years, Photo Gallery, Theory, Airmen, and more.

Free or Fee: Free.

Helicopter History Site

http://www.helis.com

e-mail: online form

RATING
✈✈

BRIEFING:

Though not a scholarly resource, this rotor wonder offers some fun pictorial history.

Whirling in a virtual fog of shaky English translation and painful layout, the Helicopter History Site emerges unscathed as an award winner nonetheless. Carefully navigating through the random ad banners and slow-to-load pictures, you'll stumble across a surprising collection of helicopter history. Your first tip is to gather ample amounts of patience and understanding.

Setting aside for the moment any hope of design or typo-free description, navigating within the site is actually easy. Thanks to a main-menu content index by timeline or company and handy continuation links, maneuverability resembles that of a Bell 430. The historical tour holds your hand while you step through the decades of manufacturers, models, and inventors. Although descriptions are brief, the pictorial reviews are good, showing a nice "helio-progression." From Leonardo Da Vinci's helical air screw to the Bell/Boeing 901 Osprey (V-22), if you're a helio-buff, raise your level of history knowledge here.

Free or Fee: Free.

TheHistoryNet Archives–Aviation and Technology

http://www.thehistorynet.com/THNarchives/AviationTechnology

e-mail: online form

Stemming from the first-rate excellence of TheHistoryNet, the Aviation and Technology Archives promise a well-documented review into the history of aviation.

BRIEFING:

Gather around the monitor for some good, old-fashioned aviation tales.

Unlike the drab presentations often associated with historical reference, TheHistoryNet's insights come alive with color and style. Obviously well-researched, the writing is flowing and interesting—a key ingredient in successful informational ventures. Photos and illustrations are many, but none so large that modem time is compromised. Each article even offers a one-page summary with a link to the full text.

For a history site covering more than just aviation, the depth of aero articles is impressive. I saturated my brain with such features as "Airmail's First Day," "The Guggenheims, Aviation Visionaries," "Kalamazoo 'Air Zoo,'" "Luftwaffe Ace Gunther Rall Remembers," and "Stealth Secrets of the F-117 Nighthawk."

When you've finally reached the end of the archived Aviation and Technology list (90+ entries), a convenient left-margin index invites you to explore other historical topics. Although these have nothing to do with aviation, the invitation still stands.

Free or Fee: Free.

US Air Force Museum

http://www.wpafb.af.mil/museum

e-mail: Patrick.Champ@maxwell.af.mil

RATING
✈✈✈

BRIEFING:

Stop by for Air Force specs o'plenty—you'll be bombarded with bombers and inundated with insight.

It's a good thing that superb site navigation, in the form of a left-margin menu, takes you by the hand. Without its directional beacon, you'd be lost in a virtual sea of bombers, trainers, and fighters. The sheer volume of images and history falls into the unbelievable category. Plan for extensive viewing if you're an Air Force admirer.

While I'm certainly not trying to diminish the importance of the actual museum in Dayton, Ohio, this promotional site is quite a spectacle in itself. In my opinion, one of the best starting points for a megasite like this is to take some virtual tours. My favorites are the Korean War, Presidential Aircraft, and R&D Hangar. Then start browsing through the multitudes of aircraft and special galleries. Modern Flight, Early Years, Air Power, Space Flight, and others present fantastic capsules of memories. From History to Engines & Weapons, there's more than a couple days worth of viewing here alone.

Free or Fee: Free.

National Air & Space Museum (NASM)

http://www.nasm.si.edu

e-mail: web@www.nasm.edu

It's simple and informative—the way a world-class museum ought to be. The Smithsonian Institute's online National Air & Space Museum gives you a virtual look at aviation and space history. Click through museum maps and exhibits, educational programs, NASM news and events, NASM resources, or just general information about the museum itself. A convenient, clickable museum map points you in the direction of your favorite exhibits. From Milestones of Flight to Rocketry & Space Flight, you'll scan through online gallery greatness.

Have a specific question or winged fancy? Just jump into the powerful search engine. Search the Smithsonian Web by typing your phrase(s) or keyword(s). It's history at your fingertips.

BRIEFING:

The Wright Brothers would be proud to take this cyber-museum tour. Although the online museum is fascinating in pixel version, don't let it quell your thirst for seeing the real thing.

Free or Fee: Free.

San Diego Aerospace Museum

http://www.aerospacemuseum.org

e-mail: info@aerospacemuseum.org

BRIEFING:

Dazzling aviation history through a futuristic Web site tour.

Wow! High-resolution pictures, well-written history, and perfect organization catapult the online edition of San Diego's Aerospace Museum into a must-see site. Aviation enthusiasts and historians are in for an exhibit tour encompassing the Dawn of Powered Flight through the Space Age.

True, online is a nautical mile from the real thing, but begin your journey here. There's hordes of fascinating information. Delve into the Montgolfier Brothers' Hot Air Balloon of 1783 (the first manned vehicle in recorded history to break the bonds of gravity), or read about your favorites: Lindbergh, Earhart, Gagarin, Armstrong, and more. Museum hours, fees, phone number, location, collection listing, and special event services are just a click away also. If you're just revisiting, quickly find new additions in What's New at the Museum, Kids Pages, and Library/Archives.

Recommendation? See the online version, and then be dazzled by the real-life stuff.

Free or Fee: Free to view, fee for in-person museum tour.

Amelia Earhart

http://www.ellensplace.net/ae_intro.html

e-mail: jellenc@ionet.net

BRIEFING:

A brilliantly orchestrated Web tribute to America's most famous aviatrix, Amelia Mary Earhart.

Delving into the wonderfully ambitious world of Amelia Earhart, this cyber-tribute justifies every one of the awards it has garnished. Expertly written, illustrated, and presented, this online tour of Ameila's courage takes the browser through The Early Years, The Celebrity, and The Last Flight. The fascinating text is easy to read and insightful. Clickable photos are scattered throughout, as well as clickable icons that take you into each chapter.

Included with Amelia's story are a few extras. Scan through unconfirmed theories about her mysterious disappearance. Or browse information regarding the Earhart Project—an investigation launched in 1988 by The International Group for Historic Aircraft Recovery (TIGHAR) to conclusively solve the mystery of Amelia's disappearance.

For those wanting to continue their education, tap into the site's film links to Flight for Freedom, Amelia Earhart, and Amelia Earhart: The Final Flight. Other related links include Discovery Gallery, Tall Cool Woman, The Sky's the Limit, Famous Women in Aviation, People's Sound Page, Mystery of Amelia Earhart, Howland Island, and The Ninety-Nines.

Free or Fee: Free.

TIGHAR

http://www.tighar.org
e-mail: info@tighar.org

BRIEFING:

Fascinating historical research masterfully displayed by the world's leading aviation archaeological foundation.

Pulled from the deepest, darkest caverns of the Web's cyber-cellars, TIGHAR (The International Group for Historic Aircraft Recovery) has itself been discovered. Not familiar with this diamond-in-the-rough? This non-profit organization happens to be the world's leading aviation archaeological foundation. Its goals of finding, saving, and preserving rare and historic aircraft are artfully displayed here—online.

Fascinating history presented through well-written and descriptive pictures is at your disposal. Read through a thought-provoking investigation into the disappearance of Amelia Earhart in The Earhart Project. Learn about the disappearance of Nungesser and Coli aboard *l'Oiseau Blanc* in Project Midnight Ghost, and probe into rumors that World War II German aircraft still survive in underground bunkers in Operation Sepulchre.

Also available: historic preservation articles, other resources, and a look into the TIGHAR Tracks Journal.

Free or Fee: Free to view. Regular, student, and corporate TIGHAR memberships available.

Dryden Research Aircraft Photo Gallery

http://www.dfrc.nasa.gov/gallery/photo/photoServer.html

e-mail: Robert.Binkley@dfrc.nasa.gov

BRIEFING:

Researching research aircraft? Or just need a couple of fun copyright-free photos? Here's your site.

A dizzying array of digitized delights is housed here at the Dryden Research Aircraft Archive home page (physically located at the NASA Dryden Flight Research Center at Edwards, California). With over 1200 images, the archive offers a huge selection of research aviation photos dating back to 1940. No copyright protection is asserted for any of the photos unless noted.

Site information is grouped by photo, movie, graphics, audio, FAQs, and many other aircraft image archives.

From the B-47 Stratojet to the F-14 Tomcat—it's all a keystroke away.

Free or Fee: Free.

The Ninety-Nines—International Organization of Women Pilots

http://www.ninety-nines.org

e-mail: webmaster@ninety-nines.org

RATING
✈✈

BRIEFING:

Cyber-headquarters for the women-only Ninety-Nines.

Founded in 1929 by 99 licensed women pilots, the Ninety-Nines organization catapults its promotion of flight fellowship with this thoroughly informative online presence. Although its crude design and lackluster site navigation offer a minor downside, the benefits are well worth the visit.

A bookmarkable must for all female aviation enthusiasts, The Ninety-Nines pages provide unending resources. In addition to organization and membership information, other topics include The Ninety-Nines in Aviation History, Women in Aviation History, Forest of Friendship, 1929 Invitation, Women Pilots Today, Hear a Little History, The Ninety-Nines Museum of Women Pilots, Learn to Fly, Future Women Pilot Program, Scholarships, Grants & Awards, Calendar of Events, Aerospace Education, Air Races, Airmarking, and more.

Free or Fee: Free unless you join.

Women in Aviation International (WAI)

http://www.wiai.org

e-mail: wai@infinet.com

BRIEFING:

Expanding the lofty horizons of active women pilots and wanna-bes.

Piloting a national and international cyber-course, the Women in Aviation International pages hone in on needed resources for aviation's women enthusiasts. Female site-seekers may be disappointed at first with WAI's lackluster design. However, once you begin clicking into its content, you'll feel better.

Tapping into WAI Events gives you dates, times, and places of WAI-specific events. In The News offers topical press releases. Scholarships gets into the good stuff with profitable scholarship information. Careers takes you to a current aviation job list with descriptions. And yes, Online Shopping is available for WAI necessities. Membership information, conference dates, *AFW Magazine* (a few editorial gems to peruse), a guestbook, and Web links are also at your clicking fingertips.

With the self-proclamation, "Women in Aviation is dedicated to the encouragement and advancement of women in all aviation career fields of interests," WAI and its site are great first steps for aviation-interested women.

Free or Fee: Free to view, fee-based membership.

The National Transportation Safety Board (NTSB)

http://www.ntsb.gov

e-mail: online form

RATING

BRIEFING:

An online look into the darker side of aviation—the accidents.

Although you may not expect it, a subtle array of graphic gadgetry takes flight here. Obviously, however, subjects with which the NTSB deals don't lend themselves to frivolity. Uncover the site's dry contents and you'll realize why its investigation wisdom becomes instantly worthwhile. Why? Because learning from the mistakes of other aviators is relatively painless.

If you're curious, you'll find tidbits of miscellany, such as About the NTSB, Recommendations and Accomplishments, Publications, News & Events, Job Opportunities, Search, and related sites.

Hopefully, though, your real investigative interests will guide you to the insights and descriptions of over 44,000 aviation accidents—there's even a handy database search capability! Or clicking into aviation accident statistics reveals tables from the annual aviation accidents press release, passenger fatality accident tables, and most recent monthly statistics. And when you're ready to learn more, lists of accident reports and studies may be ordered.

If an accident should happen to you, information on NTSB reporting is found easily online.

Free or Fee: Free.

National Aeronautics and Space Administration (NASA)

http://www.nasa.gov

e-mail: comments@www.hq.nasa.gov

BRIEFING:

Blast off into this bookmark-bound space voyage—it's quite a ride.

Building award-winning Web sites for aerospace surfers isn't rocket science. Then again, compared with the thousands of rejected sites I've hastily passed up, maybe it is.

As if the online mission control team were holding your hand through each page and topic, site navigation is effortless. Brilliant description-oriented links combine with fancy quick-click icons to launch you into unearthly euphoria. Uncluttered page layouts give you fascinating information, helpful pictures, and plenty of easy-reading "white space."

Skipping directly into the Aeronautics Technology area, you'll climb into juicy program topics such as Advanced Space Transportation, Aviation Operations Systems, Propulsion & Power, Vehicle Systems Technology, Flight Research, and more. For other NASA favorites, click into Today@NASA (breaking news and project details), NASA for Kids, Gallery (video, audio, and still images), and Space Science (planetary exploration, astronomy, and research into the origins of life).

Free or Fee: Free.

Air Force Link

http://www.af.mil
e-mail: online form

Briefing:

Official U.S. Air Force site proudly serves its online viewers with regimented excellence.

A model of Web perfection, the Air Force Link easily busts through my scale's five-plane ceiling. Criteria I've set aside for site organization, content, graphics, and intended audience are adhered to masterfully with style and purpose.

The opening menu gives you a hint as to the Web wonders you'll find lurking inside. Loading speed and appropriate presentation are perfectly balanced. However, before you get too dazzled, scan the top and side menus for your favorite site areas. Going a step further into navigational bliss, a site-searching tool is also located on the page. With the Air Force Link's resources, you just can't get lost.

Content-wise, a fantastic array of goodies begins with News, Career, Library, USAF Sites, Video, Audio, Letters, and FAQs. Specifically, Air Force Career subcategories include civilian, enlisted, officer, and retiree—all come complete with pay charts. And the Gallery page serves up an enormous list of high-resolution pictures.

The Air Force Link successfully defends its surfers against the ghastly mess so often found in aviation Web offerings. Click in and see for yourself.

Free or Fee: Free.

Cessna Pilots Association (CPA)

http://www.cessna.org

e-mail: info@cessna.org

It's no surprise that one of the world's most popular flying machines has its own association. Better still, the CPA now provides its members and nonmembers with this outstanding online technical information center.

BRIEFING:

Compilation of Cessna-specific technical insights mainly geared to the CPA member.

Mainly a gentle nudge into CPA membership, the site does a good job of describing each available Cessna-specific resource before you click. *CPA Magazine* talks about the monthly publication—devoted 100 percent to solid technical data about maintaining and operating Cessnas. TechNotes and Handouts store a series of helpful documents ranging from FAQs concerning Cessna problems to sources of discount parts. Cessna owners also will appreciate the Technical Hot-Line, Technical Library, Cessna Buyer's Guides, Systems & Procedures Courses, and Group Aircraft Insurance Program.

Updates are handled regularly with a thorough information expansion in the members-only area, with an online version of the latest issue of *CPA Magazine,* access to CPA's entire library of TechNotes, and online discussions of Cessna-related issues.

Of course, as with all organizations, online membership is clickably convenient. If you're a Cessna owner, it's a "no brainer."

Free or Fee: Free, but the good stuff requires membership in CPA.

Helicopter Association International (HAI)

http://www.rotor.com

e-mail: see site for appropriate e-mail contact

Briefing:

Your online chopper choice, courtesy of the Helicopter Association International.

Non-fixed-wing fans need only arrive at this cyber-pad and clear some room on the bookmark list. You'll be back again.

Neatly arranged in columnar text-based hyperlinks, the lengthy left-frame menu guides rotor surfers to a limitless supply of site wonders. Heliport searching (under HAI Interactive) is simply a matter of entering a city, state, or heliport. The Heli-Expo information gives you a heads up on shows near you. And the HELicopter Parts Search (H.E.L.P.S.) database teams parts listers and parts searchers effortlessly and for free!

Packed into the HAI site is a content load that would make an Aerospatiale Lamb 315B buckle. Click for industry news and archived information updated daily, online publications, chat rooms, bulletins, hot spots, education, conferences and jobs, e-mail list services, regulatory issues, aircraft for sale/lease, and way more than I have room to list.

Among my site favorites is the Cornerstone feature. This computerized civilian/military registry identifies standouts in the helicopter industry. Read through biographical data and scan pictures of thousands of helicopter professionals. It's a great way to track down a friend or colleague.

Free or Fee: Free parts searching for all!

AirLifeLine

http://www.airlifeline.org

e-mail: staff@airlifeline.org

RATING

✈✈

BRIEFING:

Find the online cyber-scoop for those in and around medical missions.

Just as the World Wide Web offers up a 24-hour labyrinth of resources, AirLifeLine proudly stands by with its own important network. As outlined in its online presence, AirLifeLine is a "nonprofit charitable organization of private pilots who donate their time, skills, aircraft, and fuel to fly medical missions."

The site clearly identifies organization goals and a myriad of FAQs with a simple contents list. Take a quick jump into About Us, Patients, Heathcare Professionals, Volunteer Pilots, and Supporters. You'll quickly notice that AirLifeLine skips the fancy GIFs and graphic fanfare associated with most of today's Web sites. The obvious concentration leans toward organization information—both for potential pilots and for those needing medical transport. For example, private pilots may learn more about volunteering by clicking into Volunteer Pilots. Here, you'll browse careful text-based details about the missions—who qualifies, how they are requested, and how each is accepted. Pilot's Liability, Requirements for Joining, and Membership Application Details are also at the ready.

Curious about signing up? Read the letters to AirLifeLine. You simply can't help but be inspired to fire up your trusty flying machine and make a difference.

Free or Fee: Free.

EAA AirVenture

http://www.fly-in.org

e-mail: webmaster@eaa.org

Briefing:

Oshkosh online—officially brought to you by the Experimental Aircraft Association (EAA).

Wittman Regional Airport in Oshkosh, Wisconsin. Perhaps you've heard of it. Every year a few aviation enthusiasts stop by for a week and mingle with airplanes and fellow aviators. Actually, the event numbers speak more clearly to its worldwide appeal: around 750,000 attendees, 12,000 airplanes, 2800 show aircraft, 700 exhibitors, and 500 forums. Numbers like these require hordes of promotion. Enter Fly-In.org.

Peek through this virtual window of Fly-In.org to get the complete skinny on this year's EAA AirVenture Oshkosh Convention. Although most of the site information during my review alluded to the upcoming event at the time, you'll get the same scoop for the next show and a thorough recap of the last one. Click your way through AirVenture Photos, Air Show Photos, General Information, Fast Facts, News, Awards, Events, and much more.

Probably during your visit you'll spend the most time boning up on the upcoming Oshkosh information. An organized display prompts you easily through the photos and descriptions of Oshkosh arrival procedures, tentative schedule of performers, exhibitors, day-by-day information, where to stay, transportation, monthly updates, forums, and admission rates.

Free or Fee: Free.

American Institute of Aeronautics and Astronautics (AIAA)

http://www.aiaa.org

e-mail: webmaster@aiaa.org

RATING
✈✈✈

BRIEFING:

The AIAA creates the epitome of Web excellence for aerospace resources.

Consistent with its professional nature and world-recognized resources, the AIAA blasts off into cyberspace with over 65 years of excellence. Technically touching on aerospace support services, the AIAA site closes in on the perfect online template for resource sites. In every category I hold dear (content, layout, functionality, and audience), this aviation presence dominates.

First, join me in a functionality analysis. The introductory page loads quickly and gives the aerospace enthusiast an entire site summary at a glance. Click either the illustrated icon or the text link—in both cases you arrive at, and transition through, your destination frustration-free. Second, layout steals a page from some Web expert's manual. You'll wander through appropriately illustrated icons and perfectly proportioned text pages. Bandwidth-hogging pictures are nonexistent, and descriptions sparkle with clarity. Third, the overall intended audience includes not only AIAA members but also *all* enthusiasts involved in the arts, sciences, and technology of aerospace.

Finally, content should be reason enough for a bookmark. Stretch your mouse hand and follow me into Career Planning and Placement Services (mainly for members), AIAA Bulletin (industry news, services, events, employment services, and more), Conferences & Professional Development, Technical Activities, Publications, Customer Service, and more.

Free or Fee: Free.

Blue Angels

http://www.blueangels.navy.mil

e-mail: bawebmaster@ncts.navy.mil

RATING
✈✈

BRIEFING:

A dizzying aerobatic spectacle in real life and online.

Blasting onto the scene in a spectacular blue streak, the Navy's Blue Angels page simply dazzles. Layout, clickable menu graphics, and well-chosen photos begin the online adventure. Then informative text tells the rest of the story. Get the initial briefing by clicking into the squadron history, covering aircraft types, missions, and objectives. Next, a look into biographies, officers, and the enlisted team focuses on the professionals behind the glitz—demonstration pilots, C-130 pilots, support officers, and the maintenance and support team.

By far my favorite site feature is the proud display of photos and movies in the gallery. You'll be bombarded with maneuvers, jets, and formations— enough to keep you busy for awhile.

Free or Fee: Free.

USAF Thunderbirds

http://www.aero-web.org/events/perform/tb/tb.htm

e-mail: none

RATING

✈✈

BRIEFING:

A fast, sleek, and thoroughly entertaining site almost mirroring the real-life stuff.

Also graceful in the online skies, the Thunderbirds display perfect pageantry and digitized delights. And, as you would expect, uniformity and solid organization provide the glue to hold it all together.

A bit light on photos, the site focuses more on history, technique, and the people behind the scenes. The well-written descriptions offer a fresh perspective and more insight anyway. Fascinating Thunderbird information can be found quickly by clicking into these sections: The Thunderbird Legend Lives On, Thunderbird History, The F-16 Fighting Falcon, Pratt & Whitney F100-PW-220 Turbofan Engine, Pilots Display American Airpower, Maintainers Keep Thunderbirds Airborne, Support Critical Piece of Thunderbird Puzzle, and Quotes.

As you'd expect in this promotional site, the current airshow schedule is conveniently clickable. Courtesy of the Aviation Enthusiast Corner (also an award-winning site in this book), you'll be hyperlinked to a long list of shows— complete with date, show title, and city. Further links provide actual locations, contacts, and performers.

Free or Fee: Free.

National Business Aircraft Association (NBAA)

http://www.nbaa.org

e-mail: webmaster@nbaa.org

BRIEFING:

An organization with a resourceful site committed to NBAA membership.

Almost needing a suit and tie to view, NBAA's site exudes professionalism, style, and a hard sell toward membership. The nonmember area offers inexhaustible information on products and services. Clickable subjects include Membership information (get the sign-up details here), Convention (dates and times), Industry Data, Government Affairs, and Products and Services.

Existing NBAA member? Just fill out the application form and get member-only access. Tap into many database-driven resources (maintenance, operations, airports), NBAA Air Mail, political and tax issues, sitewide search, and Ask the NBAA Staff.

Free or Fee: Some information is free to nonmembers. Join the NBAA and get into more good stuff.

Aircraft Owners & Pilots Association (AOPA)

http://www.aopa.org

e-mail: aopahq@aopa.org

BRIEFING:

Yes, it's a recruiting tool. However, if you're interested in joining the world's largest general aviation advocacy organization, you can sign up here.

From magazines to trade shows and fly-ins to their Air Safety Foundation, AOPA isn't used to taking the number two position. Just like everything AOPA does, this site included, skillful organization has propelled this necessary association into the aviation limelight since 1939. This site, developed for membership and club benefits information, makes it easy to see why AOPA ranks among the 100 largest membership organizations in the United States. You'll find current fly-in information, pilot news, an introduction to learning to fly, a sample of the printed magazine (*AOPA Pilot*), and yes, membership information (fees, application, etc.)

Existing members may tap into their own section for 24-hour access to searchable databases, weather information, back issues of *AOPA Pilot*, events, and more.

Free or Fee: Free to view, fee for AOPA membership.

National Aeronautic Association (NAA)

http://www.naa-usa.org

e-mail: naa@naa-usa.org

RATING

BRIEFING:

Yet another worthwhile site promoting a worthwhile club. Have a look—it may be for you.

Looking to entice new members into aviation and air sports, NAA's site invites you to grab your GPS and explore a "world of aviation" through NAA membership. Sign-up information, a mission statement, corporate membership, affiliates, contacts, and a look at NAA Today are all here. Fun little tidbits include aeronautical records, aero clubs, and air sports associations.

Not easily found in most aviation Web site link lists, NAA's special links to air sports include Academy of Model Aeronautics, Balloon Federation of America, Experimental Aircraft Association, Helicopter Club of America, International Aerobatics Club, Soaring Society of America, United States Hang Gliding Association, U.S. Parachute Association, and the United States Ultralight Association.

Free or Fee: Free to view, fees for NAA membership.

Experimental Aircraft Association (EAA)

http://www.eaa.org

e-mail: webmaster@eaa.org

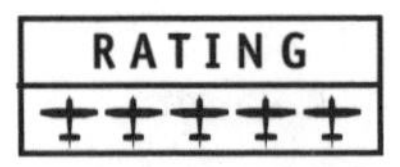

Briefing:

Here's an Oshkosh junkie's fix for the whole year.

Wish the Oshkosh fly-in lasted year-round? Before and after the real thing, join EAA members at the official EAA home page to satiate your hunger for fun flying. This slick, handsomely designed gem just shines with plenty of well-organized topics and news in many clickable areas. Reach into the left- and right-margin lists for unlimited resources such as the AirVenture Oshkosh program, all kinds of EAA-related services and products, museums, Kid Stuff (Young Eagles), and more. Go shopping in EAA's Aeronautica Gift Shop, and check out the hot topics of today summarized on the main page.

Great site and organization for enthusiasts of every age and interest—"pilots, designers, builders, dreamers, and doers."

Free or Fee: Free to view, fee for membership. Yes, there's a members-only section. Worthy information awaits.

Soaring Society of America (SSA)

http://www.ssa.org

e-mail: info@ssa.org

BRIEFING:

An online meeting place for Soaring Society members as well as nonmembers.

You've seen them in their tiny bullet-shaped cockpits floating on elongated wings. For those who glide, the thrill of soaring is infectious. The remedy is to join other thermal-hungry friends online. The official pages of the Soaring Society of America are found here—giving you a link to all phases of gliding nationally, as well as internationally.

The information contained here is organized nicely, graciously avoiding gratuitous photos. You'll find information on becoming a pilot, where to fly, contests, records, soaring services, SSA membership information, a member locator, government affairs, a soaring magazine, and safety information.

Soaring buffs and wanna-bes should look into membership. Judging by this site's organization, you'll be in good hands.

Free or Fee: Free to view, fee for membership.

Federal Aviation Administration (FAA)

http://www.faa.gov

e-mail: see site for appropriate e-mail contact

Briefing:

This information-rich governmental site will dazzle you with its efficiency.

All government jokes aside (too many to list here), one can only stare slack-jawed at this wondrously efficient, visually appealing FAA piece of mastery. When the weight of federal bureaucracies begins to creep into your life, you'll find the light at tunnel's end here.

Granted, there's some administration stuff here that most will care less about. However, slip past it into the main menu topics of Aviation Support & Regulation, For Passengers, Newsroom, Safety, and more. Not into clicking through menus? Just search the site using a nicely designed search engine to breeze through traditional red tape. You'll get your questions answered.

Or examine how to do business with the FAA, review a long list of FAA-related organizations, and read about aviation education and jobs.

Free or Fee: Free.

Office of Airline Information

http://www.bts.gov/oai

e-mail: webmaster@bts.gov

BRIEFING:

An award-winning governmental (yes, governmental) site from the Bureau of Transportation Statistics (BTS) that's shamefully useful.

Yet again, rather than rearing its ugly head of red tape, your government has chosen a course of surprising functionality and automation. Congratulations to the BTS for assembling this amazingly useful treasure of vital transportation information.

Tap into the BTS's Office of Airline Information and you'll open the statistical file cabinets housing the FAA Statistical Handbook of Aviation, BTS Transportation Indicators, U.S./International Air Passenger and Freight Stats, and the Sources of Air Carrier Aviation Data. Although some stats may be a few years old, there's a wealth of information here on airline on-time reports, passenger and freight counts, aircraft accidents, airport activity, general aviation aircraft information, aeronautical production, U.S. civil airpersonnel, and more.

Free or Fee: Free.

Air Transport Association (ATA)

http://www.air-transport.org

e-mail: ata@air-transport.org

RATING

BRIEFING:

When not embroiled in fare wars and cutthroat competition, your favorite commercial carriers come here to join forces, data, and knowledge.

Leveled off and cruising at commercial flight levels, you'll find the ATA's official Web site. This cyber-resource is always on time with 24 hours of ATA information for members and a smattering of information gems for non-ATA types. Mainly serving its 28 commercial carriers (USAirways, United, Northwest, Southwest, TWA, etc.), ATA's well-organized site features an absence of unnecessary pictures and slow-to-load graphics. A slick opening menu gets you into areas concerning general information, services, and member-only stuff. Don't shy away if you're a nonmember. There's a series of interesting things here for you too. The Airline Handbook launches into the history of aviation, deregulation, airline economics, how aircraft fly, the future of aviation, and more. You'll find industry stats, ATA publications to order, air travel survey results, and more.

Members can tap into an ATA Calendar, Airworthiness Directives, Memos and Data from the ATA's Engineering Maintenance and Materiel Council, Technical Support System, Events, Publications, and ATA News.

While we all have opinions on the efficiency of commercial carriers, you'll find ATA's site timely, useful, and convenient.

Free or Fee: Free with restricted membership area. See ATA's pages for details.

Angel Flight

http://www.angelflight.org

e-mail: webmaster@angelflight.org

RATING
✈✈✈

BRIEFING:

A nonprofit organization that not only moves you but also transports those less fortunate with medical problems to a treatment destination.

Don't expect fancy graphics, gratuitous plane pictures, or rambling commentary here. The text-only descriptions and Press Gallery are quite enough to send your hope for the human race soaring. The Angel Flight site promotes and explains the aviation community's volunteer service of getting needy folks to diagnosis or treatment. Angel Flight's volunteer pilots and nonpilots join forces to shuttle cancer patients to chemotherapy and surgery, carry people with kidney problems to obtain dialysis or transplants, and bring those with heart-related problems closer to treatment.

With this informative site, prospective volunteers will find many answers to common membership questions, including Who Belongs to Angel Flight? Who Does Angel Flight Transport? Where Do Calls Come From? Who Pays for the Flights? How Do I Join? What Happens after I Join? and, What Is My Liability?

Angel Flight's pages are simply a persuasive call to action. Where do I sign?

Free or Fee: Free.

The Mechanic Home Page

http://www.the-mechanic.com
e-mail: mechanic@the-mechanic.com

RATING
+++

BRIEFING:
A quality resource for the true aviation maintenance pro—breathtakingly bookmarkable.

If you're a person who tinkers in the cowling or wrenches on undercarriage, wipe off the grease and reach for a mouse. Settling into The Mechanic Home Page reveals a heart-pounding depository of aircraft maintenance technician information. At the center of this site is the Aircraft Mechanics Fraternal Association (AMFA)—a craft-oriented independent aviation union. Cleanly organized with great mechanic information, everything here is easily accessible. From several linked information pages to downloadable files to Java pull-down menus, every site tool is at the ready. Sift through airline news (broken down by major carrier acronyms), enter or read comments on the bulletin board, read about the AMFA, download many important FAA files, scan the news archives, read observations from industry professionals, tap into employment opportunities, learn from miscellaneous mechanic articles, and view the many hot topics currently displayed on the introductory page. The functional design and excellent variety of downloadable resources complete the package of perfection.

If you're remotely involved in the nuts and bolts side of aviation, break out your bookmark.

Free or Fee: Free.

Seaplane Pilots Association

http://www.seaplanes.org

e-mail: webmaster@seaplanes.org

RATING

✈✈✈

BRIEFING:

Have a seaplane fancy? Get your feet and your floats wet here.

Although fairly unusual in the Web world, Seaplane Pilots Association is mostly informational. Yes, that means mostly you'll find water-flying facts.

And if the lack of junky ads weren't reason enough to visit, Seaplane Pilots Association is organized efficiently too. You'll wonder why all aviation sites haven't copied its source code. Simple page links with nice descriptions get you into more water-flying fancy than one should be allowed in a Web visit.

Clickable contents include Schools, Instructors, Study Material, Ratings Info, and Upcoming Events. Further informational sections point you in the direction of briefs, links, and online catalog. It's a one-stop seaplane shop.

Members get "special" Web access to even more information resources for water-flying fans such as a rental directory, operators directory, float manufacturers, seaplane datasheets, and a Make & Model Directory and an Articles Database (from *Water Flying* magazine).

Free or Fee: Free for lots of information, but membership (for a fee) gives you even more online resources.

MicroWINGS

http://www.microwings.com

e-mail: fltsim@microwings.com

RATING

✈✈✈

BRIEFING:

Informative online hangout for flight simulator types.

Yes, I've even uncovered an association for flight simulator buffs. Here's a reliable, helpful buddy to extract the maximum simulation exhilaration from your favorite flight software. As demonstrated by this site, MicroWINGS fiercely devotes itself to all types of aerospace simulation.

For real, approved flight training or just a chance to dip into the fantasy slipstream, I've found your land-based cockpit pros. Newly updated as of review time, the site has a zillion products to review in the Global Flight Simulation Online Catalog, product reviews, download files, flight simulator crossword puzzles, events, online chatting, and a strong push toward membership in the International Association for Aerospace Simulations. With membership, though, comes hefty product discounts, a full-color magazine subscription, free software, simulator bulletin board access, and more.

Free or Fee: Free to view, but if you're into simulators, get the fee-based membership.

Bookmarkable Listings

World League of Air Traffic Controllers
http://www.wlatc.com
e-mail: riger@atchome.com
The World League of Air Traffic Controllers' site for news, chat, and contact information.

Vietnam Helicopter Flight Crew Network
http://www.vhfcn.org
e-mail: webmaster@vhfcn.org
A forum for recreational communications among aircrew members who served in Vietnam.

Lindbergh Foundation
http://www.lindberghfoundation.org
e-mail: info@lindberghfoundation.org
Informational source concerning the Charles A. and Anne Morrow Lindbergh Foundation.

United States Parachute Association (USPA)
http://www.uspa.org
e-mail: communications@uspa.org
Serving the only national skydiving association, the USPA, composed of over 34,000 members.

American Bonanza Society
http://www.bonanza.org
e-mail: online form
Informational site dedicated to the owners of Beechcraft Bonanza, Baron, and TravelAir aircraft.

Air Force Association (AFA)
http://www.afa.org
e-mail: service@afa.org
Nonprofit civilian organization promoting the importance of Air Force resources.

Civil Air Patrol
http://www.capnhq.gov
e-mail: webmaster@capnhq.gov
Online services and membership center for the Civil Air Patrol.

390th Memorial Museum Online
http://www.390th.org
e-mail: the390th@aol.com
Memorial museum preserving the proud heritage of the 390th Bombardment Group.

Rhinebeck Aerodrome Museum
http://www.oldrhinebeck.org
e-mail: info@oldrhinebeck.org
Sneak peek into this living museum of antique aviation.

The Spruce Goose
http://www.sprucegoose.org
e-mail: none
Official page delves into the history of the *Spruce Goose* with photos, historical details, perspectives, and more.

American Airpower Heritage Museum
http://www.airpowermuseum.org
e-mail: keopp@sgi.com
Official pictorial tour and information relating to the American Airpower Heritage Museum of the Confederate Air Force.

Aviation Institute
http://cid.unomaha.edu/~unoai/aviation.html
e-mail: unoai@unomaha.edu
Online recruiting tool for the Aviation Institute.

Center for Advanced Aviation System Development
http://www.caasd.org
e-mail: caasdweb@mitre.org
An FAA-funded, not-for-profit organization researching important aviation topics.

National Air Traffic Controllers Association
http://www.natcavoice.org
e-mail: Staff@natcavoice.org
National Air Traffic Controllers Association members have a bit more to say than just "traffic at your eleven o'clock position."

Aviation Safety Connection
http://www.aviation.org
e-mail: safety@aviation.org
Nonprofit site furthering air safety through discussion groups.

The Flying Doctors
http://www.flyingdocs.org
e-mail: online form
Los Medicos Voladores, or Flying Doctors, was founded in 1974 to provide health services and education to the people of northern Mexico.

North West Aerospace Alliance
http://www.aerospace.co.uk
e-mail: info@aerospace.co.uk
The alliance is an industry-led initiative dedicated to promoting the region's world-class aerospace and high-technology engineering base.

Lima Lima Flight Team
http://www.limalima.com
e-mail: Jrip@limalima.com
Online introduction to The Lima Lima Flight Team—the world's only six-aircraft civilian formation aerobatics flight team!

Vietnam Helicopter Crew Member Association
http://www.vhcma.org
e-mail: n3tef@home.com
Get in touch with Vietnam helicopter crew members.

National Air Disaster Alliance/Foundation
http://www.planesafe.org
e-mail: see Web site for appropriate e-mail
Nonprofit organization raising the standard of airline safety, security, and survivability.

League of World War I Aviation Historians
http://www.overthefront.com
e-mail: ziggurat@gte.net
Online collection of articles and information from *Over The Front,* the leading journal in the field of Great War military aviation research.

General Aviation Manufacturers Association (GAMA)
http://www.generalaviation.org
e-mail: online form
Founded in 1970, GAMA is a trade organization that represents over 50 American manufacturers of fixed-wing aircraft, engines, avionics, and components.

Regional Airline Association (RAA)
http://www.raa.org
e-mail: carol_jewell@dc.sba.com
Web-based information on the RAA, which represents U.S. regional airlines and the suppliers of products and services that support the industry.

Air Transport Action Group (ATAG)
http://www.atag.org
e-mail: information@atag.org
Web introduction to ATAG, an independent coalition of air transport organizations pressing for economically beneficial aviation capacity improvements in an environmentally responsible manner.

American Association of Airport Executives (AAAE)
http://www.airportnet.org
e-mail: webmaster@airportnet.org
Information-rich online hub for the AAAE.

Professional Aviation Maintenance Association (PAMA)
http://www.pama.org
e-mail: hq@pama.org
Well-designed site serving members of PAMA and the aviation maintenance professional.

Mercy Medical Airlift
http://www.mercymedical.org
e-mail: mercymedical@erols.com
Perfectly organized information devoted to charitable air transportation in situations of compelling need.

Angel Flight America
http://www.angelflightamerica.org
e-mail: angelflightamerica@erols.com
Information on assistance with charitable long-distance medical air transportation.

Weather

Weather to Fly

http://www.weathertofly.com

e-mail: loretta@adventurep.com

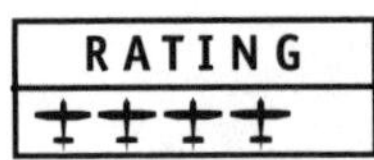

BRIEFING:

Excellent collection of weather links that's worth a visit and a bookmark.

A bit of a sideline site from your friends at Adventure Productions (extreme sports video production company), Weather to Fly tracks down an unusually large collection of weather resources for those influenced by weather. It does take a few minutes to get acquainted with the main page. Spend some time getting to know your weather options.

Site navigation is nothing more complicated than a grouping of text links for weather-related browsing, new site additions, and a navigation menu that's based on the main Adventure Productions site. Scroll down past the weather company logos and you'll uncover some excellent practical resources. Click into the main categories of Jet Stream and Wind, Satellite Images, Surface and Altitude Analysis, and Local Conditions and Forecasts. Within these categories you'll find specific forecasts and images for winds and temperatures aloft, a windcast graphic of predicted surface winds, worldwide satellite images, surface analysis (current), local forecasts, and more.

Free or Fee: Free.

PilotWeather.com

http://www.pilotweather.com
e-mail: none available

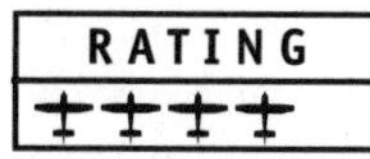

BRIEFING:
Pilot-specific weather from the talented prognosticators at AccuWeather.

Free or Fee:
Some free pilot weather information, but the site is fee-related for the premium stuff. Lots of packages and possibilities. Click in and find out for yourself. As of review time, the site does offer a free 30-day trial.

The next logical step for the world-famous AccuWeather (also an award winner in this book) was to customize its weather offerings for us pilot types. Yes, there's a subscription fee for the "premium" weather information (after a 30-day *free* trial), but don't turn the page just yet.

The new PilotWeather service features a new design and easier navigation to give you complete access to hundreds of aviation weather products and informational tidbits. With PilotWeather, you get a stockpile of real-time NEXRAD radar products, Route Weather Briefing, Flight Plan Filing with the FAA, as well as NOTAMs, SIGMETs, AIREPs, and PIREPs. Literally hundreds of weather maps and visible and infrared satellite images cater to those of us who are constantly weather curious.

Specifically, this well-designed, easily navigable site splits into a free basic service area and premium service (fee-related). The basic service avoids the skeletal information tease and actually provides some good pilot weather information, such as a route weather briefing; raw, decoded, and plain-language terminal observations; TAFs; flight rules; a severe weather section; and more. However, move into the premium possibilities and get direct FAA flight plan filing, the most current NEXRAD Doppler radars available, hour-by-hour weather, and more.

Live Weather Images

http://www.weatherimages.org
e-mail: mark@weatherimages.org

RATING

BRIEFING:

Grass-roots collection of the best in weather resources.

When I stumble across a site like Live Weather Images, it reminds me why I began scouting out the Web's best in Aviation in the first place. In the earlier years of Web site creation, when the Web was just really starting to take hold, you'd find quite a few sites like this one. Rough in design, filled with unedited and unbridled information. The fact is, though, that what you'll find here is a passion for weather and a dedication to get you in touch with the best stuff.

The author, a meteorology major at North Carolina State University, maintains this Web site as a hobby in his spare time (as if anyone has spare time in college). His goal reads "to provide free, to the general public, a concise and user-friendly weather Web site that pulls together the most valuable and frequently accessed weather data on the Internet." In my opinion, he's already succeeded. Just scan the menu to see what I mean. Main Weather Page (current weather images), Interactive Page (calculate sunrise/sunset times, heat index, wind chill, and more), U.S. Data by State, Weathercams, and Weather Humor are just a taste. Visit Live Weather Images for the rest.

Free or Fee:
Thankfully free.

Excite Weather

http://www.excite.com/weather

e-mail: online form

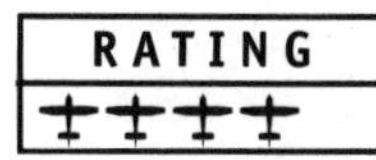

BRIEFING:

This worthy weather resource is but a tiny subsection of the vast information empire at Excite.com.

Seemingly simple at first, your weather options continue to grow as you click through Excite Weather. Obviously drawing on Excite's domination in all things Web-related, general weather topics abound here. As you'd expect, the presentation is flawless. Menus are simple. And downloading is close to instant.

When weather curiosity turns to action, start with any number of clickable selections at Excite Weather. The introduction page loads up with a simple, left-margin Weather Directory for My Weather (personalized weather), Maps, Airport Delays, Ski Reports, Almanac, Meteorologist (kind of a weekly Q&A), and Trivia. Or even without relying on any menus, just type your zip code or search by any city for current weather information. Sunrise/sunset and moon phase even load on startup.

Oh, and if you haven't already, sign up as a member of Excite to personalize your own Excite startup page. It's free and easy, and a personalized view of your favorite cities' weather will pop up each time you load the page.

Free or Fee: Free, even for Excite membership.

Weather.org

http://www.weather.org

e-mail: webmaster@weather.org

RATING

✈✈✈

BRIEFING:

Current weather images from the world's top sources at your fingertips.

On clicking into Weather.org, I'd suggest heading straight to the aviation section. However, that's your call. Certainly other specialties exist too, such as Marine, Agriculture, Global Warming, Space Weather, Climate History, Volcano Watch, Tides & Currents, Ski Conditions, and more. Me? I was more interested in the aviation side of things.

Thus, assuming you'll follow my flight path enroute to aviation weather at Weather.org, here's the preview. The linked resources are top-notch. No, Weather.org doesn't forecast or report the weather on its own. Rather, this site serves as your connection to current aviation weather data, linking up with organizations like the National Weather Service, AWC International Flight Folder, Unisys, Aviation Digital Data Page, and more.

Once at the site's aviation section, you'll be presented with the opportunity to view lots of weather graphics and depictions, such as U.S. Radar, U.S. Surface, U.S. Temperature, AIRMETs, and SIGMETs. Fog/Low Clouds, Aircraft Icing, Turbulence, Microburst, and Volcanic Ash also have corresponding images worthy of your attention.

Free or Fee: Free.

FlightBrief

http://www.flightbrief.com

e-mail: support@flightbrief.com

BRIEFING:

Your flying forecast: always clear and unobstructed with FlightBrief.

The sister site to Weather Concepts (also a five-plane mention in this book), FlightBrief seeks to stir the aviator's passion for precise flying-related weather specifics.

With flawless design and organization, FlightBrief mirrors its Weather Concepts twin with every weather facet, including some added aviation-only bells and whistles. FlightBrief pushes past its sister site with a focus on aviation and flight planning information. Lots of stuff is at your eagerly trembling fingertips. Tap into decoded (plain language) TAFs and PIREPs, in addition to METARs. There's a cool density altitude calculation program that calculates the current density altitude for airports that report weather conditions. A weather depiction table helps you decode the current weather observations at airports, puts the data into a tabular format, and categorizes the airports into VFR, MVFR, and IFR conditions. And a preliminary flight plan performs a simple calculation of "great circle" distance and average heading between airports.

Looking for a further push into subscription? How about 5-minute composite radar (now that's fast)? Hey, give the free 14-day trial a go. You've got nothing to lose and thorough weather knowledge to gain.

Free or Fee: Fee-based, but aviators will find it worthy of a 14-day free trial.

Aviation Weather

http://www.aviationweather.com

e-mail: weather@weathersite.com

RATING
✈✈✈

BRIEFING:

Free weather euphoria. Everything's here but the fancy exterior.

Will weather wonders never cease? Loads of free weather tidbits shower the moderately designed pages of Aviation Weather. No, it's not pretty, but the data are here all right. Slightly annoying ad banners give way to a long list of worthwhile SIGMETs, AIRMETs, forecasts, and specific information.

It's true that you can pretty much get similar weather details with any number of sites—some free, some not. However, Aviation Weather does a fine job at collecting a few more essentials all in one place (you'll just need to scroll to locate them). Running down the list of options, we pilots will enjoy SIGMETs (convective, domestic, and international), AIRMETs, METARs, forecasts (TAF, winds aloft, area forecasts, mountain wave, aviation icing maps, and TWEB routes), and other information in the form of time-temp-wind.

Beginners and those needing a refresher will appreciate a thorough explanation of important aviation weather topics, such as AIRMETs, convective SIGMETs, SIGMETs, international SIGMETs, TAF forecasts, and winds-aloft plots.

By the way, should you wish to wander into other weather wonderlands, simply visit the omnipresent top menu. You'll be transported to specific weather resources for marine, Canadian, agriculture, and more.

Free or Fee: Free.

Pilot Weather Briefing

http://members.aol.com/rlattery/pilot.htm

e-mail: vortex100@aol.com

RATING

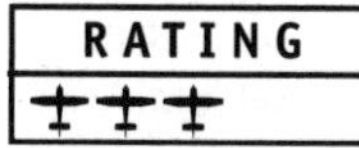

BRIEFING:

Aviation weather heavyweights pull together a complete briefing picture.

Rounding up data from a variety of respectable sources, Pilot Weather Briefing disseminates the flying weather picture clearly and thoughtfully. It's obvious much flying time was sacrificed to present such a truly useful weather resource.

Design, organization, and helpful tips are excellent. Menus are anywhere and everywhere. Fancy "lit" or "unlit" buttons tell you when you've arrived at a data page. And omnipresent satellite or radar buttons provide easy access to these valuable data. What's more, a helpful page on tips for using Pilot Weather Briefing takes you by the hand and gives you a prop start.

Surface Weather, Upper Winds, Thunderstorms, Turbulence, Icing, and Other Hazards each launch you into specific maps and analyses. Get 12- and 24-hour surface forecasts, 12- and 24-hour significant weather prognoses, temperature contours, weather depiction, AIRMETs for IFR, and more.

Did I mention that everything's free? Yes, with the help of some Web friends, Pilot Weather Briefing keeps everything current and cost-free.

Free or Fee: Free.

WeatherTAP

http://www.weathertap.com

e-mail: webmaster@weathertap.com

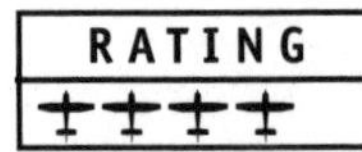

BRIEFING:

Information-rich weather wonder giving nearly real-time weather for aviators.

Still fumbling around for weather? Slip on your raincoat, pay a tiny monthly fee, and get fast, accurate weather 24 hours a day. WeatherTAP gives you left-margin "quick buttons" into local weather, NEXRAD radar, national weather, aviation weather DTC DUAT, and some other useful services. Updating every 6 minutes (the time it takes a NEXRAD radar antenna to make a complete sweep of the sky), WeatherTAP uses the exact images and information provided to the FAA's air route traffic control centers.

Of course, you'll probably dally most in Aviation Weather. Get current stuff such as National Weather Service plots, composite moisture stability, observed winds and temperatures, surface prognosis, and live data. Also worthy of your time are a host of text products and a customized route briefing section.

When you're ready to file, move directly into DUAT. The convenient link is always at the ready for the "official" briefing.

Free or Fee: Nominal fee required.

EarthWatch–Weather on Demand

http://www.earthwatch.com

e-mail: webmaster@earthwatch.com

RATING

BRIEFING:

Weather imagery at its finest from industry veterans. Bookmark this one for the 3D visuals!

Now that's where I want to get my weather—from a company famous for developing patented software that integrates 3D weather visualization with a global database to create a virtual world. EarthWatch Communications brings you cool weather visuals with cutting-edge imagery. Not convinced? Click any one of the choices on the main menu button bar.

Weather Headlines, StormWatch, Forecast Center, Satellite & Radar, Current Conditions, and EarthWatch Products lead you into pure weather euphoria. Current Conditions moves you into an updated series of U.S. weather maps—temperature, wind chill, radar, and satellite. Each is clickable for more of a pinpointed view. Clicking into the Satellite & Radar area, however, undoubtedly will serve up the most fun. High-resolution satellite imagery displays uncanny, cutting-edge visuals. My favorite? Try the U.S. 3D satellite views. Just click your region: South Central, North Central, Southwest, Northwest, and Southeast. Stunning 3D cloud layers can be seen in fairly good detail. Or radar revelers need only click once to switch to the radar image version. It's cleverly cool!

Free or Fee: Free.

World Meteorology Organization (WMO)

http://www.wmo.ch

e-mail: webmaster@wmo.ch

RATING

++

BRIEFING:

World Meteorological Organization serves as a respected jumping off point to featured weather wonders.

The World Meteorological Organization. If it seems regal and authoritative, it is. In fact, the WMO is the Geneva-based, 185-member organization within the United Nations that provides the scientific voice on the state and behavior of the earth's atmosphere and climate. Sound like a good start for world weather link searching?

Thoughtfully providing an English, Spanish, French, and no-frames version, WMO's online presence acts as a wise weather wonderland. Description of the organization itself is lengthy, with a careful emphasis on each of the WMO's "majore programmes." Learn about World Weather Watch, World Climate Programme, Atmospheric Research and Environmental Programme, and more. Of more immediate interest to us aviators, however, may be the list of worldly links. The menu of options currently (as of review time) includes libraries, meteorological sites, information sources, and specialized sites. Moving further into Additional Meteorological Information uncovers an index of meteorological information pointing to sites with national and international reports, forecasts, weather maps, and satellite images.

Free or Fee: Free.

The Weather Underground

http://www.wunderground.com

e-mail: info@wunderground.com

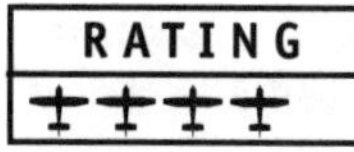

BRIEFING:

It's not a complete aviation weather provider, but it's sure a simple synopsis.

Most weather Webmasters tucked away in a dark cubicle somewhere have decided that you'll happily wait for high-byte graphics and giant radar maps to load. I, however, fall into the unhappy category when superfluous graphics and unwanted miscellany rain on my weather inquiries.

If you're an impatient person too, go underground in the Weather Underground. It's amazingly easy on your modem and efficiently organized. Don't believe me? The first screen to load gives you three options to find your weather: (1) type your city and state into a search window, (2) click anywhere on the U.S. map, or (3) find your state's hyperlink and click. It's so fast that I become giddy with weather euphoria.

Once you've arrived at the desired city, a table of current conditions provides time of report, temperature, humidity, wind, pressure, conditions, sunrise, sunset, and moon phase. Forecasts include descriptions and temperature for your desired city, as well as your state's extended forecast.

Free or Fee: Free.

USA Today–Aviation Weather

http://www.usatoday.com/weather/wpilots0.htm

e-mail: online form

RATING
✈✈✈

BRIEFING:

The big-name Web source that conveniently narrows its weather focus for pilots.

It's really your call whether or not you'd like to scan the online news from *USA Today*. However, I'd like to point you toward the site's specific weather resource for pilots, aptly named Aviation Weather. Mainly it's a collection of links specific to pilots and weather-related issues. The page does an excellent job describing the appropriate resources and offering a corresponding hyperlink. Unlike most online news sources of this caliber, graphics, pictures, and charts are nonexistent. It's simply a quick way into the weather and related tidbits you need.

At the time of review, the site makes it easy to tap into Studying Clear Air Turbulence, Lightning Protection, Understanding Density Altitude, Flying into Hurricanes, Pilots Report Hazards to NASA, NASA's Aviation Human Factors Research, and Online Weather Calculator (converting temperatures and calculating density altitude).

Perhaps a bit more specific to weather, the following topics catapult you into a great collection of associated links: thunderstorms, icing, live weather, ground school, and more.

Free or Fee: Free.

National Weather Service

http://www.nws.noaa.gov

e-mail: w-nws.webmaster@noaa.gov

RATING

BRIEFING:

Weather service professionals give you a 24-hour option for getting the real scoop on Mother Nature's intentions.

Partially obscured among the countless weather resources found on the National Weather Service's site, you'll find a healthy grouping of aviation products. Get current information from Terminal Aerodrome Forecasts, aviation weather discussions, aviation METAR reports, Terminal Forecasts, and more. Mostly text-based, the available forecasts and observations offer accuracy and speed. You won't be waiting for maps or graphics to load.

Pulling back from the focus on aviation, the National Weather Service site also thoroughly covers weather-related topics in a more general sense. When you've got some extra hangar time, be sure to scan through the Interactive Weather Information Network (warnings, zone, state, forecasts), black and white weather maps, U.S. weather bulletins, tropical cyclone warnings and products, fire weather, and Alaska products.

It may not necessarily be pretty, but all your weather is here from those who know.

Free or Fee: Free.

Aviation Weather Center

http://www.awc-kc.noaa.gov

e-mail: webawc@awc.kc.noaa.gov

RATING

✈✈✈

BRIEFING:

Preempt your local weather reporter with forecasts from the source.

Weather information doesn't get much closer to the source than this. Without pomp and pageantry, the National Weather Service and the National Oceanic and Atmospheric Administration serves up the Aviation Weather Center with an information-rich presentation. Be forewarned, though, that you won't come across colorful maps and pretty graphics. Make sure that you're up to speed on coded weather information, and you'll breeze through its text-only forecast reports.

Up to 24,400 feet, U.S. forecasts mainly include warnings of flight hazards, such as turbulence, icing, low clouds, and reduced visibility. Above 24,000 feet, the Aviation Weather Center provides warnings of wind shear, thunderstorms, turbulence, icing, and volcanic ash.

Site navigation is relatively easy with many menus above and below. While it may take awhile to become comfortable sifting through site contents, you'll eventually get to AIRMETs, Area Aviation Forecasts, Domestic SIGMETs, Terminal Aerodrome Forecasts, TWEB Routes, Winds Aloft Forecasts, and more.

Free or Fee: Free.

The Online Meteorology Guide

http://ww2010.atmos.uiuc.edu/(Gh)/guides/mtr/home.rxml

e-mail: ww2010@atmos.uiuc.edu

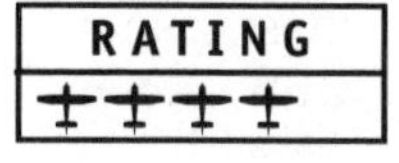

BRIEFING:

Meteorology introduction with a flair for current weather from your educated friends at the University of Illinois.

Always keeping a watchful eye on the weather and Web wonders, I'm elated when I uncover a combination of the two as well done as WW2010—The Online Meteorology Guide. Sure you can get current weather here (check out the latest handsome revision), but the real impact stems from a combination of current weather *and* instructional modules.

Meteorology instructional modules delve into a variety of fascinating topics using charts, graphics, and easy-to-understand description. Light & Optics introduces how light interacts with atmospheric particles. Clouds & Precipitation introduces cloud classifications and developing precipitation. The Forces & Wind module discusses the forces that influence airflow. And although this is just a taste of the additional contents, read on about air masses, fronts, weather forecasting, severe storms, and hurricanes.

Thoughtful navigational features are almost too abundant to list. However, it's worth noting some of my favorites: an option between full graphics or text-based site layout, a helper menu explaining site navigation, color-coded highlights for current location, and friendly left-margin menus throughout.

Free or Fee: Free.

CNN Weathernet

http://www.cnn.com/WEATHER

e-mail: online form

RATING

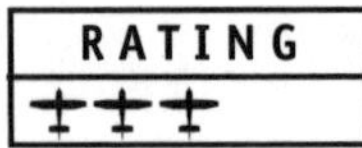

BRIEFING:

Get well-done weather 24 hours a day with CNN's interactive presence.

Adding to a swelling list of lofty weather sites, CNN's weather page blows onto the cyber-scene with its own worldly prognostication. As is often the case with supersites of this caliber, visually appealing graphics and orchestrated presentation subtly make you feel warm and cozy.

The main page quickly launches you into your desired 5-day forecast. Just type your zip code or select your state from a pick list. If world weather is what you're after, just select a region and click. Up pops a corresponding 4-day forecast, complete with highs/lows, precipitation, winds, pressure, and humidity. Current weather news, storm center, allergy report, and Biz Traveler are also standing by.

Curious about the bigger picture? Tap into your regional satellite and radar maps—they're easy to read and fast to load. Or scan through worldwide weather maps and images. From the Africa satellite image to the Europe forecast map, global weather's here for the clicking.

Although completely non-aviation-related, a current news topic list follows you in the left margin. With a click you'll jump to CNN's current news on sports, travel, world, United States, local, and more.

Free or Fee: Free.

Weather Concepts

http://www.weatherconcepts.com

e-mail: support@weatherconcepts.com

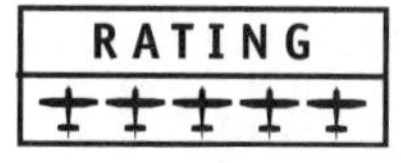

BRIEFING:

Any more great features and this weather wonder could soon replace your local TV weather personality.

Functional design and great organization make Weather Concept's weather site one of your first stops for comprehensive weather. Even if you're new to the online weather scene, the transition is easy, with an extensive explanation in each section's help section. Once you're comfortable and have given in to a 14-day free trial, have a look at AvCast, Current, Forecast, Radar, Satellite, Lightning, and more.

Pay a nominal monthly fee and delve into full-blown weather euphoria, including NEXRAD Doppler radar. The best part? Extensive aviation-related information with instant access to the identical text briefing used by flight service stations.

With area forecasts, METARs, TAFs, TWEBs, SIGMETs, NOTAMs, AIRMETs, winds aloft, and PIREPs, what more could you possibly ask for?

Free or Fee: Fee-based. Tempt yourself with a 14-day free trial.

AccuWeather

http://www.accuweather.com

e-mail: online form

BRIEFING:

This award-winning weather site makes it easy to see why good design, layout, and organization speak volumes in online communication.

From the moment you enter, there's no question that some graphics guru got ahold of these pages. Visual wizardry takes the form of icons, illustrations, and easy-to-see maps. More important, however, you'll enjoy future return trips due to careful organization. It's so easy to navigate with the masterful visual references.

From the AccuWeather introduction page, you'll be invited to enter your city (zip or city/state). Free for anyone, AccuWeather's basic service includes 5-day forecasts, links to international forecasts, local NEXRAD Doppler radar, and more. Premium service subscribers will receive real-time access to everything. AccuWeather's products include weather for virtually any spot on earth, special aviation weather section (including direct fight filing with the FAA), satellite pictures, NEXRAD Doppler radar, weather maps of current forecasts, temperature-band maps, current conditions, weather discussions, and more.

It's an organized look at weather, whether you subscribe or not.

Free or Fee: Free for some weather samples; fee-based for full-blown, real-time weather products.

Aviation Model Forecasts

http://weather.unisys.com/aviation

e-mail: none provided

BRIEFING:

An educated predictor of aviation weather.

When talk among pilots turns to weather, things usually get more serious. Abandoning any hint of whimsical buffoonery, the Aviation Model Forecasts site sticks to the topic at hand without cracking so much as a smile. From the get-go you'll plunge into technical weather data and forecasting models. Here you'll tap into colorful contour plots for weather forecasts.

Mostly designed for those on a meteorologist's level of understanding, this site does provide nonmeteorologists with the ability to view the data in varying degrees of complexity. The plots usually are updated once every 12 hours and offer a full range of forecasting—from 12 hours to 10 days. The index includes individual plot summaries; general forecast plots; initial analyses; 12-, 24-, 36-, 48-, and 60-hour and 3-day forecasts.

Yes, it may take awhile to decipher, but you're looking at some highly reliable forecasting here.

Free or Fee: Free.

The Weather Channel

http://www.weather.com
e-mail: lifestyles@talk2weather.com

BRIEFING:

The Weather Channel's masterful weather magicians mix visual delights with practicality.

The Weather Channel site—a great example of what happens when you mix talented designers, high-end software gadgetry, efficient hardware, and a team of marketing professionals. Let this group loose on the cyber-world, and you get online brilliance. Although you may not get all the pieces of your aviation weather data here, you will get quick, accurate, searchable weather tidbits, including jet stream information, winds aloft, air turbulence, and a national airport overview.

If you're a weather enthusiast with a hankering for facts and trivia, you've found your utopia. Start by typing in your zip code or city and "Go!" Instantly, you'll get stuff like current temperature, wind speed/direction, relative humidity, 7-day forecasts, barometric pressure, and current conditions.

After you've checked out all your favorite cities, don your slickers and splash through the other fun items here. Dip into Breaking Weather, Tropical Update, International Cities Forecast, Maps, Atmospheres, Exotic Destinations, and Interact (boards/chat, photo gallery, and photo of the week). All the clickable topics move you into logically presented information with visually captivating graphics—with an option to customize your own page.

With this expert site, there's just no downside—unless the predicted weather keeps you grounded, that is.

Free or Fee: Free.

WeatherNet

http://cirrus.sprl.umich.edu/wxnet

e-mail: macd@cirrus.sprl.umich.edu

BRIEFING:

Whether you're ready or not for over 380 weather sites—it's here, it's unbelievable, and it's waiting for you.

Sure, weather's important to all of us aviation types. However, what you've got here is weather obsession—in the most positive sense, that is. Luckily, the weather-worshipping folks at WeatherNet have assembled easy access to over 380 North American weather sites. Yes, over 380. Get your bookmark list cleaned up, and get ready to add. At review time, the 380 sites are simply listed in alphabetical order (a tad unwieldy), but look for better organization coming soon.

Aside from this unprecedented library of weather links, you'll find the usual forecast prompted by your city/state/zip code/country and more great WeatherNet features. Highlights include Forecasts (organized by state), Radar and Satellite (cool clickable U.S. map), Weather Cams (local photographic peeks into selected cities and popular resort destinations), Travel Cities Weather, and more.

So, if you've got a few minutes to dabble with Doppler or see through satellites, get out your umbrella—you'll always run into weather here.

Free or Fee: Free.

Intellicast

http://www.intellicast.com
e-mail: online form

RATING
+++

BRIEFING:
Knock-your-socks-off graphics combine with user-friendly organization to form weather magic.

Yes, there seems to be a weather site around every link and search engine, but Intellicast has created a unique and well-organized look at worldwide weather. Marvelous graphics and useful visual delights reign supreme here.

There's hordes of cool maps, weather-related icons, and menus. Get started with a main menu that leads to USA Weather, International Weather, Travel, and Skiing. Topic-specific weather even includes forecasts for golf, tropical getaways, the great outdoors, and sailing. Searching by city is easy. There's a map of popular cities, as well as a quick zip code or city search box to get to your favorite city's weather. Once you find your city of choice, you'll become weather savvy with instant information relating to temperatures and forecasts and a host of images to view. The long list of weather images includes radar, a radar summary, satellite, NEXRAD, and precipitation.

While you're visiting, don't forget to check into the monthly almanac, featured site selections, and Ask Dr. Dewpoint. It's fun weather fancy for everyone.

Free or Fee: Free.

Bookmarkable Listings

Real-Time Weather Data
http://www.rap.ucar.edu/weather
e-mail: gthompsn@ucar.edu
Weather data categorized by satellite, radar, surface, upper air, aviation, and more.

Charles Boley's Weather
http://www.cwbol.com
e-mail: cwbol@hiwaay.net
Personal collection of described links to current weather maps, radar images, and weather newsgroups.

Atmosphere Calculator
http://members.aol.com/nywx/atmoscal.htm
e-mail: nywx@aol.com
Calculates dewpoint, relative humidity, wind chill, heat index, and more based on information given.

National Climatic Data Center (NCDC)
http://www.ncdc.noaa.gov
e-mail: webmaster@ncdc.noaa.gov
Online collection of resources from the NCDC—the world's largest active archive of weather data.

Sunrise/Sunset
http://tycho.usno.navy.mil/srss.html
e-mail: online form
Automatically calculates sunrise/sunset, twilight, and moonrise/moonset with given longitude and latitude.

Pilot Resources

Crew Email

http://www.crewemail.com
e-mail: info@crewemail.com

RATING
✈✈✈

BRIEFING:
A crew member version of Hotmail brings pilots and crew an e-mail-friendly choice.

Is it just me, or does everyone enjoy a simple, intuitive, easy-to-use Web tool now and then? Crew Email is one of those nice little services that can really make life easier. Basically, it's a free e-mail service for crew members. You've heard of Hotmail, right? No? Here's the briefing: Hotmail is a free e-mail service and is based on open Web-based technology. The Web browser, which is more universally available than any proprietary e-mail program, is used for e-mail. Web-based e-mail (through services like Hotmail and Crew Email) is accessible worldwide, offering a cross-platform e-mail solution. And it's even easier to use than traditional e-mail programs.

Unlike Hotmail, however, Crew Email is designed specifically for flight crews around the world to access their private e-mail or calendar from any computer that has a local connection to the World Wide Web. The increased availability of Internet kiosks and public Internet connections in hotels, libraries, cafés, airports, and universities makes remote access sites pretty plentiful.

Now that you're all jazzed on this cool e-mail service, be sure to scan a few other site features during your Web stay at Crew Email. In addition to free e-mail, you'll find both private and public calendars, Crew Chat, Track a Flight (no matter how many times I see this, it's still fun), a reminder service, and a Send-a-Postcard service.

Free or Fee: Free.

Aerospace Online

http://www.aerospaceonline.com
e-mail: info@aerospaceonline.com

BRIEFING:

The undisputed online water cooler hangout for aviation maintenance and avionics pros.

Here's a fine example of an expertly produced aviation catchall site that could fit under just about any category in this book. Why categorize it under "Pilot Resources"? The reason really comes down to variety. Pilots and aviation staff will find Aerospace Online to be a thoughtful tool for technical, regulatory, operational, management, and product information about the aviation industry. It's emphasis, though, is on issues of interest to the aviation maintenance and avionics communities in air transport, business and corporate, rotorcraft, and military aviation.

Loaded with text links, descriptions, and stellar design, Aerospace Online shows off its "e-zine" presentation on the home page. The first impression may be a bit intimidating. There are industry news headlines, a Product Center (Buyer's Guide, Product Showcase, Request a Quote), a Professional Center (Job Search, Recruiter Center, Employer Spotlight, Training, and Resources), and more. So, yes, it's overwhelming. But that's why the powers that be created the bookmark. Just come back often.

Scroll through the site to find online catalogs, Deal of the Day, and Editor's Choices (feature products). Oh, and don't miss the Download Library. Handy software and tools are only a click away (some are even free!).

Free or Fee: Free.

Aircrewlayover.com

http://www.aircrewlayover.com

e-mail: HeyYou@aircrewlayover.com

RATING

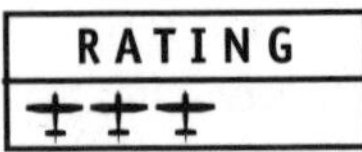

BRIEFING:

Proverbial pit stop for inquisitive crew during downtime.

Climb into the right seat with me as we venture into the off-duty time of aviation professionals. From the moment you hear the conductor yell, "board" (not sure why), you'll be invited into a friendly family of fellow crew members.

Developed just for you, the professional aviation crew member, a nice repository of services and entertainment awaits for your leisure downtime. You'll tap into insights regarding favorite restaurants, activities, points of interest, shopping areas, side trips, and affordable hotels worldwide. To search, just find your corresponding city. Or enter your own suggestions for fellow crew members to view. Speaking of viewing, you may want to check into the photo area for a collection of user-submitted photos. Whatever you do on your layover or in your aircraft, Aircrewlayover.com wants to know.

Lots of aviation links with nice descriptions and many book-buying suggestions are at the ready in case you've researched every city on the site. But that's unlikely. There's much to see.

Free or Fee: Free.

Aircraftfuel.com

http://www.aircraftfuel.com
e-mail: info@aircraftfuel.com

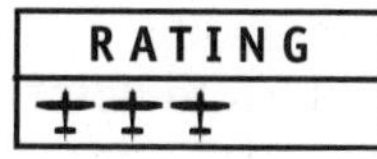

BRIEFING:

Track down fuel prices throughout the United States before you go wheels up.

Universally on the minds of all pilots, whether private, commercial, corporate, or otherwise, fuel prices can be difficult to predict, with prices that also can be hard to swallow. Finding fuel at prices we can accept is a necessity. Welcome to Aircraftfuel.com. This site does the legwork; you reap the benefits of the company's research.

This handsome Web site interface introduces you to your options of viewing fuel prices, membership to look up fuel prices (free and instant), FBO membership signup, FAQs (brief, but necessary), and a list of contact e-mail addresses. Search for your fuel by state and then city. A long list of clickable cities is the entry point to specific FBO fuel information. In tabular form, you'll see the airport ID, the FBO name, phone, fax, supplier (Chevron, AVFUEL, etc.), prices, and types of credit cards accepted.

Although my run around the Aircraftfuel.com pattern was a little shaky, I was still fond of the site enough to add a bookmark. The initial turbulence stemmed from navigation that had me using the "Back" button and reentering my user name and password often. It'll probably be fixed by the time you stop by.

Free or Fee: Free.

Air Partner

http://www.airpartner.com

e-mail: charters@airpartner.com

RATING

++

BRIEFING:

Self-described as the "world's largest corporate aircraft broker," Air Partner does excel online.

Ahh, the world of aircraft charter. Jet-setting, puddle-jumping, or globe-trotting. Getting from point A to point B in noncommercial glory certainly has its appeal. For many reasons. If you're serious about air charter, anywhere worldwide, you might begin your research here.

Air Partner maintains its status as the "world's largest corporate aircraft broker" (the company's words, not mine). With operations in London, New York, Fort Lauderdale, Paris, Cologne, and Zurich, the company obviously has some pretty good coverage and connections. Just click on the Who We Are, Capabilities, and AP Worldwide to get a better idea of Air Partner. Me? Of course, I go straight to the gallery of aircraft. Frankly, the aircraft selection is enormous. From an eight-seat Piper Chieftain to a superluxurious B747 Executive, your options are really limitless.

Noted site functions include an inquiry form, a price guide for a wide range of aircraft, a self-contained intra-airline support program, and an urgent freight charter division. While I was eager to view the Air Partner video online, something was a bit glitchy during my visit. You may have better luck.

Free or Fee: Free, unless you decide to charter something.

FltPlan.com

http://www.FltPlan.com

e-mail: Support@FltPlan.com

BRIEFING:

Excellent online flight planning resource concentrates on function, not style.

Okay, I admit that *free* usually gets my attention. Not that I'm cheap, mind you. I'm just always looking for aviation resources that will benefit fellow flyers. And, *free* usually gets the attention of fellow flyers. So, already up in the points standing is FltPlan.com because, yes, it offers a cool, *free* service.

FltPlan.com, geared toward the general aviation world, assists in the creation of professional IFR flight plans and navigational logs for serious corporate, charter, or business pilots. Pilots also can use the site for planning a flight and the Navigation Log enroute. FltPlan.com provides IFR routing, winds aloft, aircraft performance, an airport database, frequencies, FBO information, recommended alternates, and more.

FltPlan.com's Web site navigation is relatively painless. Enter your user name and password to get things going. Lots of instruction, FAQs, and helpful hints get you through if you're a first-time user. I do recommend a quick click into the Introduction, Limitations, Overview & Tutorial section. It just might give you a prop start into the site's updates and offerings. Although site design is a little lackluster, it's really the functionality that takes center stage. FltPlan.com seems to handle the input forms, processing, and output well, and that's what counts here.

Free or Fee: Free.

Fboweb.com

http://www.fboweb.com

e-mail: green@ids.net

RATING

BRIEFING:

Worthy informational resource is packed to the rafters with stuff of interest to any pilot or FBO.

Pilots and FBOs might as well add this one to their favorites list right now. This Internet-based tool for both pilots and FBOs contains a wealth of clickable information that is handy and current. Pilots may want to obtain general information and plan their flights, whereas FBOs may want to promote their services, facilities, and local information to curious pilots.

The "e-zine" look for this resource wonder is clean and refreshing. Getting around in the site is effortless, and information is presented with perfect clarity. Skip around to the various sections, such as airports, NAVAIDs, FBSs and services, flight plans, weather, N-number search, AIM (Aeronautical Information Manual) online, links, discussion forum, and more. Curious if something's new at Fboweb.com? Just scroll down to the bottom of the introduction page to see a long list of site changes and updates (great reference for repeat visitors).

Perhaps the best feature about Fboweb.com is its cutting-edge dissemination of information via wireless handhelds. Yes, handhelds! You know, like the Palm VII. Other devices, such as the WindowsCE/PocketPC and wireless Web-enabled phones, are also forthcoming we're assured. This means that users of these devices can be "on the field" and wirelessly access airport information, get the latest METARs and TAFs, or file a flight plan. Yes, that's cool.

Free or Fee: Free.

Pilot's Web

http://www.pilotsweb.com

e-mail: editor@pilotsweb.com

RATING
✈✈

BRIEFING:

Up-and-coming article-driven resource for up-and-coming pilots.

The primary colors of blue, red, and yellow make up your first-page palette. Menus, layout, and offerings are all relatively simple. Site navigation is a bit suspect. However, my hope is that Pilot's Web builds on its currently humble content and presentation. The foundation's great. The idea is strong. How can you go wrong with a free resource with aviation-enlightening articles? Some are practical. Some are esoteric.

A closer examination of the current menu contents (as of review time) leads you to the Pilot's Web Network (aviation bulletin boards and chats) and a journal of insight into flight training, weather sense, air navigation, principles of flight, articles, and more. In my opinion, some of the better articles included "The Earth's Coordinate System," "Distance and Direction," "Time and Time Zones," "Magnetism and the Magnetic Compass," "Airfoils, Lift and Drag," and more.

Yes, there are quite a few well-written, worth-your-time articles here. However, there's room to grow, too. I look forward to the Aviation Terms Explained section (nothing there at review time) and the nonfunctioning METAR/TAF Training Program. (It'll probably be working again when you read this.) Follow the instructions to download this program for free.

Free or Fee: Free.

AirCharterSolutions

http://www.airchartersolutions.com
e-mail: info@airchartersolutions.com

RATING
✈✈✈

BRIEFING:

Empty-leg seekers may find that a charter flight has a seat for them. Check in at AirCharterSolutions to find out.

Here's something I would have expected to see more of by now. Filling those charter flight empty legs online. No, the process at AirCharterSolutions isn't completely automated. In fact, the site simply provides the means for air charter companies to list their upcoming empty legs. However, searching by day, month, year, and region *is* automated and handy. Just enter your proposed trip dates and see if you match up with a scheduled charter flight. Found a match? Just contact the respective air charter company. The rest is left up to your negotiating ability.

Aesthetically, AirCharterSolutions receives top honors. It's built for speed, beauty, *and* functionality—a lethal combination in the Web world. The design, layout, and navigation obviously were the work of real professionals. I'm sure you too will have nothing but kudos to contribute as you click around the site.

Note: It should be made clear that AirCharterSolutions is not a travel agent, broker, or air charter operator. The company simply provides the means for air charter companies to list their empty legs. Or in their own words, "AirCharter Solutions is an informational Web site designed to provide the consumer with knowledge of alternative air travel accommodations through the use of classified ads posted by various private air charter companies."

Free or Fee: Free.

Floridapilot.com

http://www.floridapilot.com
e-mail: pilot@floridapilot.com

BRIEFING:

Florida pilots rejoice. Your all-in-one aviation resource shines online.

Free or Fee: Free.

I know, I know. This site's focus is pretty limiting. Sure, Florida's a large state with lots of aviation activity. However, why the award-winning mention, you ask? In a word: perfection. If all states went to this extreme in creating an absolutely comprehensive piece of aviation Web mastery, they'd get a mention in this book too.

So now to the specifics. Floridapilot.com overflows with thoughtful resources for Florida-bound pilots. The flawless introduction page presents countless clickable sections and information tidbits right at startup. Destination Guides features four airports at review time (St. Pete/Clearwater, Daytona Beach, Key West, and Cedar Key), as well as a Central Florida Flying Guide. The Flying Safety section features a fantastic MOA & RA Refresher. Florida pilot NOTAMs, weather, restaurant guide, airport guide (searchable), classifieds, and Florida aviation news headlines also fill the home page to Floridapilot.com. Overload? Maybe to some. But thoughtful, resourceful pilots certainly will appreciate its wide Florida-only appeal.

One of my favorite site features is a constantly updated list of Florida fly-ins and events. The list is huge, with lots of quality description for each event. There's even a separate air shows table that displays at a glance about 6 months or so of air shows. Get dates, hosting airport, contact names, phone numbers, and Web address (if applicable).

HangarConditions.com

http://www.hangarconditions.com
e-mail: webmaster@hangarconditions.com

RATING
++

BRIEFING:
New hangar haven opens its doors to the online aviation world.

Right at the get-go you should realize that HangarConditions.com is relatively new and evolving even as you read this. Its content is a bit sparse but growing. So why the mention here? Quite simply, it's a necessary concept that the online aviation community has tended to ignore. Hangars. I think we can agree that they are an important ingredient in aviation. Some need them. Some don't. However, if you happen to fall into the former category, clear some room for a bookmark right now, and thank me later.

HangarConditions.com is a nationwide source for aircraft hangar information, including hangars for rent and sale, hangar manufacturers, door manufacturers, contractors, and so on. You can request a quote (listings of companies or individuals needing a quote on a building, door, etc.). Look into rental hangars and tie downs. Shop hangars for sale by state (includes size, price, and airport information). Scan the growing list of hangar manufacturers, hangar door manufacturers, hangar contractors, and airport equipment (power carts, tugs, tow bars, airport lighting, fuel trucks, and more).

Some listings in certain categories are thin, but it'll fill out in time.

Free or Fee: Free.

The Air Charter Guide (ACG)

http://www.guides.com
e-mail: online form

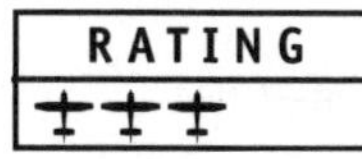

BRIEFING:

Briefing: A must-be-viewed site for charter operators and passengers.

Whether you're on the passenger side or the cockpit side of air charter, you simply must have a printed and online version of ACG at your fingertips. The printed 700-page reference edition teams up with this online guide to give you hundreds of charter operators and brokers worldwide. The index conveniently sorts charter operators by location, name, or specialty; and U.S. and international brokerage services by name and specialty. There's even a complete list of aviation services providers—from insurance to sales.

Although the printed edition tells the complete story about each operator, the online "zine" makes retrieval of basic info easy. Also on tap: charter industry info, discussions, and news. Industry reference topics give you a peek into: travel weather, city guides and maps, airline reservation systems, limousine and car rentals, and travel publications and associations.

The site is expertly organized, fast, and helpful. Chartering? Dial this one up!

Free or Fee: Free.

CrewStart

http://www.crewstart.com

e-mail: online form

RATING

✈✈✈

BRIEFING:

Crew-oriented collection of linked goodies.

Some savvy readers of this book may wonder why CrewStart landed here, under the category of "Pilot Resources" and not under "Directories." The answer, you'll find, lies in the content. Or, better said, the links to the content. CrewStart is certainly an aviation directory site. Lots of links take you anywhere aviation-related. However, with such a strong collection of crew-friendly link resources, it was simply destined to park and tie down here.

Built for speed, not beauty, the introductory link list page tells the story. Preflight your CrewStart experience with a look at the categories: search engines, CrewMail, News, Aviation Photo Search, Crew Travel Tips, Communication, Best Aviation Sites, Crew Tools, Restaurants and Hotels, Chat and Bulletin Boards, Air Data, Crew Fun, Aviation Jobs, Crew Health, Crew Dating, Sounds and Animations, and much more. Specifically fun, useful, and otherwise noteworthy destinations are Date a Pilot, Airline Nicknames, Airtoons (fantastically funny), Live ATC, and more. Just click in and find your own favorites.

Be sure to at least sample the CrewStart bulletin board CrewRumors. Read absolutely wild, wacky, and uninhibited airline-type gossip. Sample topics include rumors and news, wanna-bes and vacancies, hangar talk, zero tolerance, and more.

Free or Fee: Free.

Flightneeds.com

http://www.flightneeds.com

e-mail: support@flightneeds.com

BRIEFING:

Carefully wrapped selection of high-powered pilot and scheduler resources.

A smattering of bits, pieces, and summarized teases tempts you into Flightneeds.com membership. And why not do it? It's free after all, and painless. Sign up in about 30 seconds and get your e-mailed confirmation code. At login, launch into lots of worthwhile resources. Oh, and in case you were wondering who's involved in the site, do Air BP, Jeppesen, Aviation International News, PilotPortal.com, FlightTime.com, and AC-U-KWIK ring a bell?

Created especially for us pilots and schedulers, Flightneeds.com ties together a string of resources from some well-respected companies and serves them to us on a virtual silver platter. You'll have at your clicking fingertips everything you need to scope trips and order flight services. Now that's service!

Remember, membership is free and allows you access to fuel and trip services, ground services, a pilot's toolbox (distance calculator, unit converter, flight requirements, and more), weather, aviation news, classifieds and forums, shopping (via PilotPortal.com), charter services, and more.

Free or Fee: Free, but you'll need membership.

Pilot Medical Solutions

http://www.leftseat.com

e-mail: info@leftseat.com

RATING
++

BRIEFING:

Are chances a little shaky that you'd pass your medical? Check in here first.

At first glance, you may think I've guided you into a corporate pitch promoting prequalification services regarding your aviation medical certificate. And you're right. Partly. Yes, Pilot Medical Solutions in Tulsa, Oklahoma, has a medical protection program that prequalifies pilots before they apply for their medical certificates. However, there's much to read online, and it's free.

Concerned about the possibilities of passing your medical? Check into Pilot Medical Solutions' site for a wealth of aviation medical examination information. Although the site's riddled with navigation and design blunders, true information seekers will press on to find a gold mine of news, standards, reports, and FAQs. Specific menu options include General Info (Aviation Medical Examiner seminars, contact numbers, criteria, and statistics), FAA Standards and Forms, Special Issuance information, Medications, Statistics, FARs, and the Aviation Medical Examiner Guide (in read-only format with Adobe Acrobat).

Well, why not try it? Take the FAA Medical Practice Test. This test is designed to identify areas that may jeopardize FAA certification or that will require extensive documentation prior to approval.

Free or Fee: Lots of free information. The actual prequalification services are fee-related.

FlightTime.com

http://www.flighttime.com

e-mail: request@flighttime.com

RATING
✈✈✈

BRIEFING:

Aircraft charter made easy and as close to automatic as you can get.

Without compromising your Web time with cheesy graphics and drawn-out sales pitches, FlightTime.com reigns supreme in many cyber-areas: content, design, page navigation, and organization. Dial it up and see for yourself.

Air charter passengers, air charter operators, and frequent travelers in general need to make room on the bookmark list for this resourceful Web wonder. Get quotes and book aircraft charter flights to any destination worldwide right online. It's easy, painless (except for the bill), and convenient. Lots of help is available too, in online form and via phone.

However, it's the extras that propel this air charter arranger into award-winning status. FlightTime.com's cool CharterWizard determines suitability by analyzing the number of passengers, baggage, distances of each trip leg, runway lengths, price, and other aircraft comfort and performance features. Also indicated are the estimated times enroute for each leg and estimated local arrival times at each destination for each recommended aircraft.

Free or Fee: Free to use site. The air charter flights are *not* free.

Virtual Flight Surgeons (VFS)

http://www.aviationmedicine.com
e-mail: online form

BRIEFING:

Thoroughly useful online aeromedical consultants keep you flying wise with real answers to real questions.

Whatever ails you fellow flyers, VFS is standing by with an answer online. Visually organized to perfection, you'll avoid the ER-type panic and settle into professionally thorough aeromedical consultation.

From the get-go your board-certified physicians are a mere click and question away. A left-margin menu outlines your choices: medical information (great!), hot topics, health/nutrition, FAQs, aeromedical links, a bookstore for medical/aviation reading, medical examination tips, FAA policies, and answers to medication and drug questions.

Skipping through the fee-oriented service levels of The Waiting Room and Office Visit, you'll be bombarded with an array of specific aeromedical services, including private consultations, a premium newsletter, an FAA liaison service, expert witness testimony, and FAA medical certification issues.

What a fantastic, flowing information resource. Questions about allergies and medications, hypertension, vision surgery, vision abnormalities, cholesterol, or others? Yes. You'll get an answer. Take advantage for health's sake.

Free or Fee:
Some services are free; others are fee-related but reasonable.

Crewmembers International

http://www.crewmembers.com

e-mail: webmaster@crewmembers.com

RATING
✈✈

BRIEFING:

An insider's resource center for the high-flying crews of airline, military, and corporate aviation.

Kind of new to the commercial carrier online scene, Crewmembers International scores big with a well-conceived resource for airline pilots, flight attendants, military aviators, corporate pilots, and others.

Even if you're not among the important crew list above, you probably know someone who fits the Crewmembers description. The exchange of relevant information here is simply unparalleled Web-wide. With weak page design as the only stumbling block, you'll pull away from the jetway and be on your way with great insight nonetheless. The key? Just rely on a long-winded left-margin list of topics (mirrored at page bottom as well).

Aviation professionals will marvel at the sheer volume of tips, tricks, facts, and rumors found in a host of areas. Crewmembers are invited to transition through Crashpads, Airport Cars, Ticket Exchange, Aviation Hazards, Layover Tips, Classifieds, Reunions, Crew Message Boards, Tech Center, Track a Flight, Pilot Lounge Chat, Resumé Service, and more.

One handy aside: If appropriate, try the Find a Friend Board. The novel approach to tracking down fellow aviators seems fun and useful.

Free or Fee: Free.

Aviation Law Corporation

http://www.aviationlawcorp.com

e-mail: phil@aviationlawcorp.com

RATING

✈✈

BRIEFING:

True aviation law information online crushes the myth that all aviation law sites are just smoke and mirrors.

Free or Fee: Free, unless you get yourself into trouble and really need Mr. Kolczynski.

Phillip Kolczynski's law Web presence is surprisingly refreshing for only one reason: loads of free aviation law, hints, tips, and explanation. Yes, the focus is on credentials, experiences, and successes. In short, it's an online pitch for clients. And no, Mr. Kolczynski's not a Web page designer. His shtick is aviation law. So don't go crying over boring, "no-frills" page design.

Every online indication points to "big time" aviation law experience. Just dip into any one of a number of in-depth articles (most are old but still relevant) on a variety of topics and make up your own mind. The writing is long, thorough, insightful, and current. Even if you aren't in trouble with the law (or the FAA), take a peek at the variety of informational offerings: "Aviation Product Liability," "NTSB Investigation Guide," "Avoiding FAA Sanctions," "Post Air Crash Procedures," "FAA vs. Airmen (10 Tips)," "Victims vs. Defective Products," "Aircraft Owner Liability," and more. The experience is there, the topics are real-world, and the recommendations are genuine.

Although the articles are worth a bookmark alone, there's much more. You'll find a relatively current online newsletter covering many important topics, an aviation safety report form (with FAQs), information on preparing for trial in federal court, and some favorite aviation poems to lighten things up a bit.

Aviation Law Corporation's award-winning site uses candor and credentials to get your attention. My free advice? Listen to counsel. He or she has your best interests in mind.

RentalPlanes.com

http://www.rentalplanes.com

e-mail: webmaster@rentalplanes.com

RATING
✈✈✈

BRIEFING:

A huge fleet of rental planes is merely awaiting your command to "clear prop" and go.

True, there's a large contingent of nonowners out there (myself included) who play the rental game. Gladly it's not as awful as renting cars, but I'll welcome any aids to make the process easier. And RentalPlanes.com is just such an aid.

Just as a fully IFR-equipped Cessna 172 has the tools for guidance, RentalPlanes.com comes outfitted with top- and bottom-page navigation bars to get anywhere sitewide. The site makes it real easy to just strap in and go without a checkflight or walkaround. The menu's simple: Find an Airplane (free), Add an Airplane (also free for a limited time), Welcome to New Users (nice, brief introduction), Add/Edit Accounts, Quick Search by state, and the ever-present shameless promotions.

Just know up front that the database of worldwide planes is huge. Take advantage of the search criteria form to limit your choices. Some limiting criteria include make, model, year, power type, cruise speed, maximum rental rate, avionics, total time, ratings required, airport identifier, special characteristics, and more.

No, you probably won't uncover a Cessna 150 outfitted with a stormscope and TCAS, but you will find everything in an airworthy rental based on your reasonable perimeters.

Free or Fee: Free at review time.

Crashpads.com

http://www.crashpads.com
e-mail: staff@crashpads.com

At Crashpads, you're almost guaranteed to be pleasantly bombarded with a host of site solutions and niceties for flight crews searching for some in-between downtime.

BRIEFING:

You're getting sleepy, very sleepy, and the FAA says you're out of duty time. Get to Crashpads.

Just listed crashpads, special offers, and a huge resource center for flight crews are just what the FAA ordered. To view the enormous and secure crashpad database, however, you must become a member (verify your employment with an airline). For those listing crashpad accommodations, no membership is required, just submit the required information. Even if your accommodations are covered, crash here anyway during your ground time. An aviator's array of delightful resources is surprisingly helpful with just a few clicks. Check airline flight availability, follow an interactive flight tracker (courtesy of TheTrip.com), or get hotel phone numbers. Look up the airport codes of the world. Read about airline world news. Even check into a live, real-time chat room.

If you're in the commercial carrier business, Crashpads.com tucks you in and keeps you refreshed.

Free or Fee: Free. Membership is required to access Crashpad's database, and you must work for an airline. All the other resources don't require a membership.

Air Routing International

http://www.airrouting.com

e-mail: techsupport@airrouting.com

BRIEFING:

Get on track and stay there with a newly emerging site from a long-time air routing professional.

One of the pilot's greatest resources? Yes, that's right, it's information. Take an online trip with Air Routing International, and you'll discover why it's now an award winner.

A couple of key features are sure to become the international pilot's best friend: airport locator, pilot feedback, and the AIRMAIL newsletter. Although excellent page presentation and link organization should command some attention and review, I'd rather concentrate on these functional flight facets. Use the airport locator (by city or identifier) to retrieve location data on a huge list of airports worldwide. Even get time and distance information using a handy calculator. Just enter departure location, arrival location, and airspeed in knots (the calculator will automatically include a 15-minute bias on takeoff). The pilot feedback form, offering real experiences from international pilots, launches you into a long city-alphabetized list of hints, suggestions, and warnings. It's a huge compiling of interesting yet informative remarks.

And if ground time permits, get plenty of air routing news from around the globe from AIRMAIL by the AR Group.

Free or Fee: Free. Hints at fee-related air routing services.

Air-News

http://www.air-news.com
e-mail: none available

RATING
✈✈✈

BRIEFING:
Real opinions from world travelers shine light on the globe's best and worst.

Worldwide flyers take heart and take note, Air-News is for you. Avoid any advertised hype about eating out, hotels, and entertainment, and jump straight to real opinions from real people who speak their minds online. Air-News collects experiences and presents them using a fashionably designed database. The categories are many, and the comments are candid.

Initially, a handsome interface welcomes you aboard with an at-a-glance world-region choice to begin your journey. Select from Europe, North America, Asia, South America, Africa, Australia/New Zealand, Worldwide, or In Flight. You may even opt to go straight to the newest articles or search via a host of left-margin methods.

Whatever your searching methods, suffice it to say that finding your category and region is easy. Once you begin reading the real-life opinions, you'll be hooked. Categories within which folks offer suggestions include hotel, fun, nightlife, shopping, warnings, eating out, entertainment, travel, health/medicine, hints and warnings, and passenger matters. The entries themselves are fresh and unedited. Each record offers information for location, category, subject, remark (the opinion or suggestions), author/e-mail, and age of entry (nice!).

Free or Fee: Free.

QuickAID

http://www.quickaid.com
e-mail: online form

BRIEFING:

Invaluable airport information resource is free and fast. What more could you ask for?

Perhaps you've already tapped into the airport-related travel information dispensed by a QuickAID information kiosk found at many major airports. If not, you need only grab a mouse, dial in, and search QuickAID online. By simply following the easy-to-use system of menus, you'll quickly uncover a wealth of information about ground transportation, shops, services, area hotels, and even terminal maps.

Personally, I had no interest in company-related details (personnel profiles, company background, and contact information). If you're like me, you'll simply skip the corporate stuff and make a beeline to QuickAID's cyberheart: airport information.

The compilation of data's huge, but the site's surprisingly fast nonetheless. Begin by clicking your airport of choice (a list of airport names with identifiers is provided). Although not every airport is represented, lots of the majors are here. Get information on ground transportation (taxis, buses, shuttles, etc.), area lodging, airlines, airport services and facilities (restaurants, banks, business services, children's services, parking, and more), a terminal map, and Airport Yellow Pages.

Free or Fee: Free.

Travelocity

http://www.travelocity.com
e-mail: online form

BRIEFING:

Mega travel haven does everything but fly the plane.

While it's not impossible, getting the full grasp of Travelocity's rich information wonderland isn't a carefree endeavor. The interface, though perfectly designed for ease of use, is intimidating at first. However, fellow aviators need only gather their courage and begin clicking.

Frankly, if you're just in it for the destination guide and not the reservations stuff, your choices will be that much simpler. Certainly, with Travelocity you can do a lot of things online: book a flight, rent a car, reserve a hotel, book vacations and cruises, and more. If you were so inclined, you could spend weeks pouring over this site's saturation of travel lore.

At least at review time, I was more intrigued by the destination guide, maps, and weather. Just narrow the search to a selected city and get the scoop online. City facts, Activities, Orientation, Overview, and When to Go are written well and packed with current information (courtesy of Lonely Planet and Frommer's publications). Weather reports, including current conditions and 5-day forecasts, are available for over 740 cities. And street maps make directions a breeze.

Free or Fee: Free.

The Professional Pilot's Wait Time Web Site

http://www.pilotwait.com

e-mail: webmaster@pilotwait.com

RATING

BRIEFING:

Real recommendations about dining, lodging, and things to do from fellow flyers.

Waiting for the weather to change? Waiting for a passenger? Waiting for an engine overhaul? Whatever your wait state, The Professional Pilot's Wait Time Web Site invites you to make good use of your ground time. Maybe it's time to become a little more enlightened concerning great places to eat and stay and fun things to do worldwide. Gathered by people who travel for a living, the accumulated local knowledge provides real-life recommendations and honest observations.

The page presentation's simple. Keeping a "down to earth" theme, the offerings are friendly and understated—just right to host a free fellow pilot forum for real-life recommendations. A left-margin column gives the stats at a glance: revision date, new listings for which states, and a listing count of restaurants, places to stay, and things to do. Listings are encouraged to be of a local nature (big chain establishments are discouraged). Most recommended listings offer establishment address and phone number with an unedited description. A corresponding rating (in stars) and cost scale (in dollar bills) give you an even clearer picture of each recommendation.

As of review time, there were well over 700 listings, representing 49 states and 27 countries. Now that's substance!

Free or Fee: Free.

Fillup Flyer Fuel Finder

http://www.fillupflyer.com

e-mail: none

RATING
✈✈✈

BRIEFING:

Finding your way through this fuel finder is worthwhile.

Maybe someday Fillup Flyer Fuel Finder will find the time away from the pump to clean this confusing yet bookmarkable online presence. Don't misunderstand, Fillup Flyer is in this book for a reason. Under the thin veil of design lurks a very usable fuel resource that is worth the initial effort.

Fee-based Fillup Flyer provides members and nonmembers with fuel price reports based on routes, nonstop, multidestination, area, or statewide. What are your options for report information delivery and requests? Most choose computer, but fax, voice, and mail are also available. Once you gather your visual orientation, you'll have the major topics in sight. Click into About Fillup Flyer, Admin, General Info, Membership and Costs, Member Reports, Nonmember Reports, Premier FBOs, Sample Reports, Nationwide Fuel Station Price Statistics, and more.

Hey, if a little extra scrolling and clicking are worth saving up to a dollar per gallon on your next trip, then tough it out and assign this one to your favorites list.

Free or Fee:
Fee—sign up annually (recommended) or pay by report.

Aviation Information Resource Database

http://www.airbase1.com

e-mail: sales@airbase1.com

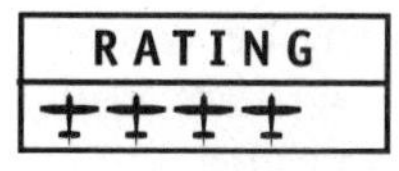

BRIEFING:

Let your mouse do the clicking through this big yellow book of online aviation resources.

Cleared for an informative flyby, you are invited to mouse your way around this resourceful labyrinth aptly tagged the Aviation Information Resource Database. Brought to all cyber-flyers free of charge by AIRbase ONE, this dominating database equates to a computerized Yellow Pages.

Flip through its topics and you'll see what I mean. Delve into over 12,000 aviation businesses listed in over 1100 categories. In addition to thousands of service-related listings, there's a complete facility directory of all public/private airports and heliports. Also at your curiosity's convenience are aircraft and engine parts, FBOs, avgas or jet fuel suppliers, and an exhaustive general aviation events calendar.

When you're really ready to pinpoint a preference, searchable subjects get you there with powerful queries. Just pick a topic: aviation businesses, airports, fuel suppliers, fuel prices, DUATS planning, lodging, restaurants, ground transportation, general aviation events calendar, and other aviation Web sites.

Although hard to do, getting lost at this airbase is corrected easily with a handy "Need Help?" button. It's a progressive taxi through an industrious airpark.

Free or Fee: Free.

Equipped to Survive

http://www.equipped.com

e-mail: dritter@equipped.org

BRIEFING:

The survival instinct is alive and well with this must-bookmark site.

Long flights over water. A perilous single-engine journey over mountainous terrain. What if disaster strikes? Will you be prepared? The fact of the matter is survival—you may have only one chance. Make it count with the online excellence of the Equipped to Survive site.

First, a note on aesthetics—don't expect any. Honestly proclaimed up front, the site's author emphasizes information, not imagery. Second, site navigation simply consists of links to topics and "Previous Page/Next Page" buttons and a drop-down menu. So nothing fancy here either. However, dip into the third criterion of content and you strike gold. The focus, you'll find, is on equipment—what is useful, what works, and what doesn't. Most of the site information is based on the author's research on wilderness and marine survival from an aviation perspective.

Fantastically insightful articles worth printing and saving include "The Survival Forum," "Basic Aviation Survival Kit," "Ditching" (for pilots), "Aviation Life Raft Reviews," "Survival for Kids," "Survival Skills and Techniques," "Aviation Life Vest Reviews," and more.

Free or Fee: Free.

TheTrip.com

http://www.thetrip.com

e-mail: feedback@thetrip.com

BRIEFING:

A travel agent, a taxi driver, a map interpreter, a hotel concierge, a maitre 'd, and an airport TelePrompTer all rolled up into an online information fest.

Straying a tad from my aviation-only focus, I must include this favorite of mine for the frequent traveler. Whether you're the pilot in command or some other MD-80 crew is getting you there, TheTrip.com becomes the ultimate in travel insurance.

Wonderfully educated design professionals weave their obvious skills and combine perfect page scripting. Thoughtful navigational bars and subtle icons guide you effortlessly through fantastic travel data. Under Flight, you'll become your own travel agent—checking flight availability and actually making reservations. There's even a real-time flight-tracking query to check any flight's status. Trip Planner helps you to book a flight, car, or hotel. And saving the best for last, the Tools for Travel category ably spews out specific weather, lodging, dining, or city-map data based on your chosen city.

Tired of business travel stress? Settle into TheTrip.com for a predeparture briefing.

Free or Fee: Free.

Air Safety Home Page

http://airsafe.com

e-mail: tcurtis@airsafe.com

BRIEFING:

Airline safety analysis gives you the hard truth and worthwhile advice for airline travel.

Odds are you were bombarded with media overload from the last airline disaster. Newspapers, radio, cable, and even Internet news join the bandwagon when heart-wrenching air carrier tragedies occur. In the interests of safety, this site's veteran airline safety analyst brings to light important industry information not easily found in one source.

Some events you may remember painfully, other airliner mishaps may have escaped your attention. In any case, the facts and observations found in the Air Safety Home Page may open your eyes to valuable passenger advice.

Complex navigational gizmos and graphics are virtually nonexistent. However, Air Safety's claim to acclaim is hard, reliable data and important advice. Read about the top 10 fatal jet airliner mishaps, fatal jet airliner events by model, the top 10 air traveler safety tips, child safety, top 10 questions about airline safety, and more.

I know it may not be the most cheerful topic, but I urge you to point your browser here for safety's sake.

Free or Fee: Free.

Air Safety Investigation Resource (ASI)

http://www.nationwide.net/~airsafety

e-mail: airsafety@dfw.net

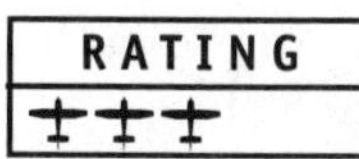

BRIEFING:

Strategic selection of quality aviation links with safety in mind.

Similar to my publication's goal of rounding up worthy sites, the Air Safety Investigation Resource seeks to narrow the infinite sea of aviation's online cache. Simply organized into a catchall table of contents, the site topics dribble down the page.

Although you'll be keeping your scroll bar arrows and "Back" button busy, this site's worth the extra hunting time. Navigation and aesthetics rank among the ho-hum variety, but the thoughtful collection of hyperlinks catapults it onto the bookmark list. Obviously, the time spent sifting through and collecting quality reference sources took priority—and rightfully so.

Check into these summarized link areas: databases of accidents, ADs, NTSB investigations, service difficulty reports, weather maps and text-based forecasts, aviation organizations, and airport information.

Until an official online aviation library takes center stage in cyberspace, this is it.

Free or Fee: Free.

EarthCam

http://www.earthcam.com

e-mail: cammaster@earthcam.com

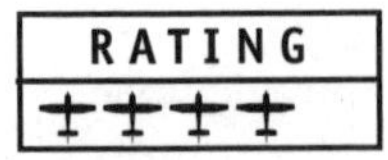

BRIEFING:

Be everywhere at once with a 24-hour peek into the world's live video cams.

Room with a view. Window on the world. An eye in the sky. Whatever your description, EarthCam *is* the source. Blossoming into a global cybercam hub, EarthCam compiles and categorizes the limitless living directory of live video cameras on the Web.

Generally used by aviation enthusiasts as a quick destination peek, these handy cams offer glimpses of live weather conditions—instantly. Although obviously not worthwhile for navigation or flight planning, the cams do give some quick insight into current conditions. Better still, the wide variety of cams isn't limited to airports or large cities. You'll get a look at the Bay Bridge traffic, Chicago's lakefront, or Disney World.

Colorful page presentation, clever icons, and convenient searching tools cheerfully guide you into an unbelievable array of cams. Peer through traffic cams, weather cams, business cams, educational cams, scenic cams, and more. Or get a bit more serious about weather with a clickable world map broken down into regional satellite views.

My favorite part? Previewing text descriptions before downloading the pictures saves hours of frustration. Be sure to read them prior to clicking—there are some time-wasters in the mix.

Free or Fee: Free.

High Mountain Flying in Ski Country U.S.A.

http://www.nw.faa.gov/ats/zdvartcc/high_mountain

e-mail: Jo.Albers@faa.gov

RATING

✈✈

BRIEFING:

A text-based summary of essential mountain flying information with specific emphasis on Colorado's ski country.

Tooling around the Rockies and other high-altitude airports means extrasharp mountain flying skills. While expertly briefing pilots on the intricacies of Colorado flying, this educational site gives all mountain-bound pilots a great heads up!

Brought to you by the concerned folks at the Denver Air Route Traffic Control Center and the FAA, this site's a great resource for explaining high mountain distinctions. Do's and Don'ts of Mountain Flying, in particular, is a well-written laundry list of hints, guidelines, and safety essentials. Print it and keep it in your flight bag—it's excellent. Winter Flying is an equally informative masterpiece reminding pilots of the hazards of winter weather and aircraft operation. And appropriately, there's a valuable refresher on density altitude.

If you're planning a flight into Colorado's ski country, check out these clickable topics: Commonly Flown Colorado Mountain Passes, Ski Country Airways Structure, Ski Country Airports, Flight Watch, Colorado Pilots Association Mountain Flying Course, and more.

Bookmark this site for safety's sake—it's a keeper.

Free or Fee: Free.

The Hundred Dollar Hamburger

http://www.100dollarhamburger.com

e-mail: jpurner@100dollarhamburger.com

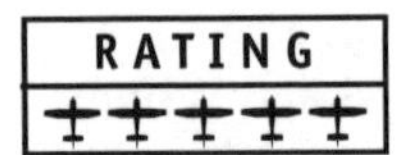

BRIEFING:

Looking for fly-in eats? This burger's well done!

Yes, I've mentioned The Hundred Dollar Hamburger before, linked to other sites. However, something this good deserves its own check ride.

Love to fly? Love to eat? Point the GPS toward The Hundred Dollar Hamburger—a Pilot's Guide to Fly-In Restaurants. Conscientiously updated, you'll get first-hand reviews and information on just about every general aviation fly-in worldwide.

First, click on your desired state or country (16 as of review). Second, choose the selected city to get an honest, straight-from-the-pilot's-mouth review. And third, view your selection, complete with reviews. Write-ups are awarded one to five burgers, with five being best.

Be responsible. Use the burger rating form and serve up a PIREP of your favorite or not-so-favorite fly-ins.

Free or Fee: Free.

DUAT.com

http://www.duat.com
e-mail: webmastr@duat.com

BRIEFING:

With free online flight plan filing and weather briefing information, this site should be every pilot's first stop.

Still one of the best resources in civil aviation today is the FAA's DUAT—Direct User Access Terminal—service. This valuable site provides current FAA weather and flight plan filing services to all certified civil pilots. The service is available 24 hours a day, seven days a week at no charge to the user—fees to operate the basic DUAT service (weather briefing, flight plan filing, encode/decode) are paid by the FAA (with your aviation tax dollars). With DUAT, you'll access completely current weather and NOTAM data. Instantly select specific types of weather briefings: local briefings; low-, intermediate-, or high-altitude briefings; and briefings with selected weather types. The DUAT computer also maintains direct access lines for flight plan filing. You can file, amend, or cancel flight plans.

Plan on filing with DUAT often? Try using MyDUAT. Save all your pilot and aircraft information in DUAT so that you don't have to retype it for each briefing! MyDUAT is your own DUAT profile, where you can store information for one or more pilots and aircraft.

Free or Fee: Free.

Flight Data

http://www.flightdata.com

e-mail: webmaster@flightdata.com

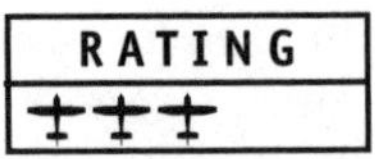

BRIEFING:

This aviation data player weaves a quality web of data-driven information.

Among the aviation data sites that spin webs of wonder, Flight Data ranks right up there, where oxygen is needed for extended periods of time. If you're a data site admirer, odds are you'll virtually hyperventilate at this site's volume, speediness, and succinct design.

A surprisingly quick-to-load list of clickables represents your introductory menu—complete with clickable main directories and subcategories. As you would expect, data ooze out of each area. Scan FAA Databases for FAA examiners, mechanics, FARs, N-numbers, and pilot data. Get pointers to instruction with Ground School's FAA exams, AC recognition, and flight schools. Or tap into the Classifieds for aircraft and employment listings. True aviation junkies might even want to delve deeper into its link lists for supplies, owners, builders, simulation, and more.

If you're not into the intensive data searching, skip over to other site niceties. Visit a few Air Traffic Control centers with the live link provided, click through a Graphs area for graphs based on 300,000 registered U.S. aircraft, or get a quick peek of your area weather.

Free or Fee: Free.

TheAviationHub

http://www.theaviationhub.com

e-mail: avhub@theaviationhub.com

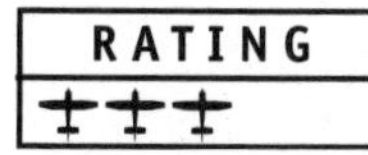

BRIEFING:

Researching general aviation information and services? Begin your prospecting here.

Founded in 1999, TheAviationHub's purpose in life is to serve up a unique take on providing service procurement, marketing resources, and information to the general aviation industry. The company offers a wide array of information and transaction services including service procurement, marketing tools, information resources, aircraft listings, performance comparisons, industry updates, and research tools. Yes, TheAviationHub represents a good place to start when you're searching for general aviation services.

The company's proprietary Service Procurement System integrates the entire service request process by bringing message communication software, an active comprehensive database, and a customer support center all together and all in one place. Cool, huh? I thought so too.

Site design and organization make online maneuvering at TheAviationHub a breeze—almost representing reason enough for a bookmark.

Free or Fee: Free.

Bookmarkable Listings

Airwise Hubpage
http://www.airwise.com
e-mail: feedback@airwise.com
An independent guide to worldwide airports and aviation/airline news.

The Homebuilt Homepage
http://www.homebuilt.org
e-mail: webmaster@homebuilt.org
Central reference to homebuilt/experimental-class aircraft.

FlightTime.com
http://www.flighttime.com
e-mail: request@flighttime.com
Worldwide aircraft charter provider with an outstanding online interface.

Flight Watch
http://www.flightwatch.com
e-mail: JBDenmanPA@cs.com
Flight Watch directs you to aviation's finest legal resources with your own online attorney.

Best AeroNet
http://www.bestaero.com
e-mail: see Web site for appropriate e-mail
Business aviation fuel network with jet fuel uplifts at over 700 sites worldwide.

Introduction to GPS Applications
http://www.redsword.com/gps
e-mail: jbeadles@pobox.com
Position your browser here to find worlds of information on global positioning systems.

F. E. Potts' Guide to Bush Flying
http://www.fepco.com/Bush_Flying.html
e-mail: none provided
Online book represents its printed predecessor with an unequaled reference for the bush pilot.

See How It Flies
http://www.monmouth.com/~jsd/fly/how
e-mail: jsd@monmouth.com
Not your standard book on flying airplanes, this new spin on piloting gives even veterans a reason to jot down a few notes.

Learning to Soar
http://acro.harvard.edu/SSA/articles/learn_soar.html
e-mail: online form
A nondescript gem of useful, GIF-less information about learning to pilot a glider.

LegFind.com
http://www.legfind.com
e-mail: nate@legfind.com
The global standard interface for charter delivering real-time data on aircraft availability worldwide.

AeroPlanner.com
http://www.aeroplanner.com
e-mail: webmaster@aeroplanner.com
Nice collection of flight-planning tools, databases, and online sectional charts.

LiveAirportInfo.com
http://www.liveairportinfo.com
e-mail: jeffgiannina@liveairportinfo.com
Excellent presentation of airport information, city Web cams, and travel guides.

Flight Training and Flight Schools

Advanced Topics in Aerodynamics

http://www.aerodyn.org

e-mail: filippon@aerodyn.org

RATING
✈✈✈

BRIEFING:

Advanced but interesting exploration of aerodynamics.

Meet Dr. Filippone, the person behind the pixels. He studies computational aerodynamics, aerodynamic design, unsteady aerodynamics, rotary aerodynamics, propulsion systems, multiphase flows, large-scale optimization problems, and racing aerodynamics.

I like a fine wine and football on the weekends. However, that doesn't make us so different that I'm not intrigued by his work and in-depth study on topics that do affect me. Advanced Topics in Aerodynamics is a comprehensive and rapidly growing Internet guide to aerodynamics, computational fluid dynamics, and related technology (aeronautical sciences, road vehicles, propulsion, energy conversion systems, design, and more). A little too "high brow" for you? Do not skip along just yet. This site's worth your time.

Expertly written and presented, the content is organized through a table of contents and Quick Guide. Once you've found something of interest, chances are you'll dive into a world of report data, tables, graphics, sketches, examples, results, and photos. Example topics include atmospheric flight, aerodynamic drag, high-lift aerodynamics, unsteady aerodynamics, low-speed aerodynamics, high-speed aerodynamics, wings for all speeds, and more.

If you are of the inquisitive variety, grab a snack and a beverage of choice, you may be here awhile.

Free or Fee: Free.

Nupilot.com

http://www.nupilot.com
e-mail: info@nupilot.com

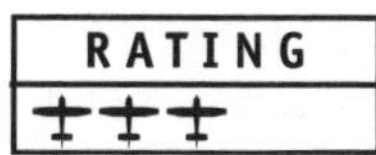

BRIEFING:

Interactive journey makes ground school learning fun.

Let me just suggest that as you first explore Nupilot.com, you should head straight to the Register area and get your password. Don't worry, it's free. Next, click into Ground Effect for an excellent "virtual ground school" into the theory of flight. It's a fantastic animated, 3D, and interactive journey for both student and private pilots.

After you have inundated yourself with Nupilot's ground school instruction topics, peruse the main home page menu. You'll notice a little something for everyone (including resources specifically for Canadian-based pilots and students). Get your feet wet, and click into Hangar Talk (aviation newsgroups and forums), Pilot Shop, a flight school database (Canadian-based), preflight forms, weather links, and more.

Probably the second best reason to visit, after the Ground Effect section, is a limitless section called Resources. Tap into stuff like enroute and terminal fee calculator, flight-planning form, aviation and daily news, air shows and conventions, the alphabet and Morse code, abbreviations and acronyms, and more.

Yes, some of the resources are really for Canadian-based pilots. But there's still plenty of reasons for *all* pilots and student pilots to add this one to their bookmark list.

Free or Fee: Free.

Start Flying Cessna

http://www.startflyingcessna.com
e-mail: online form

RATING
+++

BRIEFING:
QuickTime movies, photos, and sales pitches make you want to spread your wings with Cessna.

I know Start Flying Cessna is simply a huge promotional presence for Cessna. Even with the sales pitch reverberating in my mind, I can't help but gawk at the visual, navigational, and informational perfection that cries out for a bookmark. The site performs, entertains, and informs so well that it makes you want to run out and buy up a few Cessnas (new, of course) and dole them out to prospective pilot friends.

The clickable and expandable left-top menu guides you to the richly promotional site-wide gems. Let's Fly, Higher Learning, Pilot Reports, Take a Discovery Flight, Start Flying, and more tease you into browsing the site. And, I must say, if you do, you'll be hooked on Cessna.

No, the quantity of content is nothing to write home about. Start Flying Cessna really is a promotional tool at best. It's still worth a peek, though. And if you're close to taking the next step in getting your private pilot's license, Start Flying Cessna is a bookmark must.

Free or Fee: Free.

Free FAA Test Prep

http://freefaatestsprep.com

e-mail: webmaster@freefaatestsprep.com

RATING
✈✈

BRIEFING:

This fun little test prep is chock full of questions and extras.

This free interactive study guide serves up hard-hitting test questions for students who want to sharpen their knowledge. It's fun. It's free. It's easy. And it's bookmarkable.

Select your overall testing area (private, commercial, instrument, CFI, or ATP). Next, you'll select a particular chapter of study (such as Airports in the Private Pilot Test). Enter your name, e-mail address, and number of questions you'd like on your test. That's it. Your test pops up with multiple-choice questions. Take the test and check your answers.

Extras include the ability to view the scores and e-mails of the last 25 test takers, the opportunity to discuss questions and testing in chat room, and discussion forums.

Note: Some of the questions require you to examine charts and depictions available in printed test prep materials. Of course, the site provides the link so that you can buy online.

Free or Fee: Free.

MyWrittenExam.com

http://www.mywrittenexam.com
e-mail: info@mywrittenexam.com

BRIEFING:

Superb site serves up sample FAA written exams.

Prospective private pilots and those upgrading into the next level should feel comfort in the fact that they live in the Internet age. MyWrittenExam.com makes studying for your next test that much easier. With any of these tests, preparation is the key, and MyWrittenExam.com is the leader far and away.

MyWrittenExam.com makes practice easy with its online pages that come pretty close to Web perfection. Menus are easy to use. Personalization and user interaction features are many. And load time is lightening quick. The practice tests, made up of *all* the official FAA questions and answers, will help you ace any of the FAA written exams for recreational, private, instrument, commercial, flight instructor, flight engineer, and airline transport pilot.

The tests are timed (just like the real thing). Except here you can pause and save your progress. Stop and come back at any time. After you've completed a test, you'll get a grade in each knowledge area and a list of questions that you answered incorrectly.

Free or Fee: Free for a limited time (as of review time).

FlightSafetyBoeing

http://www.flightsafetyboeing.com

e-mail: webmaster@fsbti.com

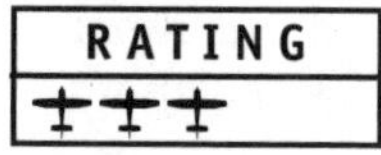

BRIEFING:

Two powerhouse aviation concerns form the basis of airline training perfection: FlightSafety Boeing.

Maybe you're not out shopping for training centers for your regional airline. Then again, you could be. The point is, although FlightSafetyBoeing online isn't for everyone, those who are in the business certainly will find it worthwhile.

Formed back in March 1997, FlightSafetyBoeing Training International provides airlines with high-quality, reliable flight and technical training delivered closer to key regional centers. This independent company draws its combined strengths (and obvious Web design talent) from its parent companies: FlightSafety International and The Boeing Company.

To fully experience the FlightSafetyBoeing adventure online, grab the pull-down navigational menu and start clicking. Main menu topics include Company Information, Flight Training, Boeing Business Jet, Technical Training, Training Materials, and eLearn. You'll uncover lots of catalogs with specific course information, training center locations and contact information, and an introduction to the eLearn system of Web-based instruction.

Free or Fee: Free.

Pan Am International Flight Academy (PAIFA)

http://www.panamacademy.com

e-mail: info@panamacademy.com

RATING

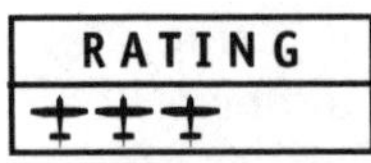

BRIEFING:

Big-name training academy serves up a jumbo portion of its training information online.

Design is good. Site navigation follows traditional patterns. Downloading is speedy. However, ask me about content included in Pan Am International Flight Academy, and I could literally go on for pages. Career aviation professionals looking for initial and recurrent training really need to include Pan Am International Flight Academy on the bookmark list.

Whatever your questions about training, this site's covered it. Probably in more than one area. Click into the main menu for your primary information-gathering adventure. Learn about the academy first with What's New (site additions and a few nice informative tours), About PAIFA, and Contact PAIFA. Next, lift off into the training program information with Career Pilot Development, Type Ratings, Business Aviation, Regional Airlines, Commercial Airlines, and Air Traffic Control.

Pages and pages of information on programs, courses, fees, and more are tucked in every corner of this online gem. Click in and find out.

Free or Fee: Free.

Student Pilot Network

http://www.studentpilot.net
e-mail: spn@studentpilot.net

BRIEFING:
Reach into this network for tips from, discussions about, and experiences of student pilots.

C'mon, we were all students once. Maybe you still are. Regardless, every seasoned and not-so-seasoned pilot can relate to the Student Pilot Network's enthusiasm, insightful forums, and first-time stories.

I think you'll agree that the presentation is flawless, adding to your visiting experience. Menus are carefully placed in nonintrusive frames. Links, icons, and site specifics are clear and easy to find. The contents of articles and page summaries are written well and nonintimidating. What a great secondary resource for those in pursuit of flight.

Obviously, Student Pilot Network endeavors to make flight school searching relatively painless. Identify schools meeting your particular needs, and make quick comparisons with the many schools in the database based on your criteria. Other site offerings include sharing experiences among pilots in the many forums, providing excellent learn-to-fly-articles offering tips and answers from aviation experts, and pointing to aviation scholarships.

My favorite feature, however, is the pilot interviews. Loads of informative Q&As will fill your screen based on type of pilot. Read about the successes of the seasoned pilots. And live the day-to-day realities of a student pilot. Whatever your skills and aspirations, Student Pilot Network is invigorating and entertaining.

Free or Fee: Free.

StudentPilot.com

http://www.studentpilot.com

e-mail: feedback@studentpilot.com

BRIEFING:

An honestly thorough student pilot resource. Free instruction, tips, and more. How can you go wrong?

When just starting to break out sectionals for the first time, beginning aviators appreciate any and all educational tools. Yes, there are always plenty of books, maps, charts, and fellow aviator tips at the prepilot's disposal. However, technology shines again with StudentPilot.com's worthy learning resource online.

Obviously not meant as a substitute for personal flight instruction or as a comprehensive reference material, StudentPilot.com's brilliance is in its ability to summarize neatly and present clearly. Just take a preflight walkaround of StudentPilot.com. Page design and organization are model examples of Web perfection. Simple menus and lack of time-consuming graphics make page piloting a breeze. Taxi easily into listed, yet well-organized articles, instruction, forums, and shopping.

Sure, page mechanics pass with high marks, but the true test of aptitude surfaces with a content checkride. The virtual flight school truly delivers a self-described "one-stop teaching tool for student pilots" with thorough descriptions of the basics. Preflighting an aircraft, navigation techniques, ground reference maneuvers, and weather report analysis are just a few. True students will enjoy a huge compilation of fellow pilot experiences—uninhibited and easy to relate to. They're inspiring, enthusiastic, and real.

Free or Fee: Free.

Spartan School of Aeronautics

http://www.spartan.edu

e-mail: awillmann@webzone.net

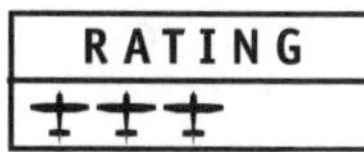

BRIEFING:

Spartan's online brochure peeks into the school's training haven for aeronautics hopefuls.

Faster than you can say "aviation maintenance and professional pilot training," the Spartan site captivates the eager aeronautically inclined to pursue their dreams. The introduction page gives you an inquiry form link on startup (the site wastes no time).

It's easy to get jazzed with so much information at every turn. Wind your mouse through a myriad of main-menu highlights, such as the Spartan Executive, the latest school news, and financial aid information. The long list of left-margin site links takes you further into your Spartan search. One of the better site aspects, especially for those contemplating an aeronautics career, is the Aviation Career Area, detailing the nature of work, working conditions, employment outlook, and qualifications and advancements for each Spartan career specialty. Learn about aviation maintenance, avionics/instruments, nondestructive testing, communication electronics, and professional pilots. Obviously, though, more than just industry information can be had at Spartan online. Tap into the course catalog, college newspaper, student housing, international enrollment, school calendar, and more.

Free or Fee: Free.

Sierra Academy of Aeronautics

http://www.sierraacademy.com

e-mail: info@sierraacademy.com

BRIEFING:

Future airplane pilots, helicopter pilots, dispatchers, and mechanics need only point their mouse here to begin.

On initial page load, quick-changing aircraft photos whet your appetite to get serious about an aviation career. Snapshots of pilots in training, singles, twins, and torn-down engines give you a visual taste of what could be.

With perfect presentation, Sierra Academy's site takes on an air of professionalism, which is just the kind of start you'll want when choosing a career-level flight school. The introductory options are simple and non-threatening. Avoiding the complex maze of online tours peppered with boring school news, Sierra Academy of Aeronautics' online vision is a bit clearer than most. Great information is at the heart of the hype for Sierra. For instance, each career path program is thoroughly detailed and thoughtfully arranged. Just call up the Airline Pilot section. You're pleasantly greeted with industry information, program particulars, and related information (financial aid, job placement assistance, and low-cost college degree options).

To begin, though, you're invited to read through a nicely succinct introduction to Sierra Academy information concerning women in aviation. Move on into your field of choice with the details of airplane pilot, helicopter pilot, aircraft dispatcher, and aircraft mechanic.

Free or Fee: Free.

College of Aeronautics

http://www.aero.edu

e-mail: mariez@aero.edu

RATING

BRIEFING:

Prep for the college in your pj's. Get the scoop on the college's aeronautics programs.

With 80 percent of College of Aeronautics graduates employed with aviation industry giants, the college's promotional Web site almost demands respect. However, even if you don't know the school's stats, its online presentation is inviting nevertheless. With a host of possible clicking directions, the site introduces you to its information and resources via lots of rollover-type menus.

There's no question the navigation's good. Text menus and buttons get you anywhere site-wide. Dial in the GPS on the mouse pad and depart. You simply can't get lost. Opening into an aeronautical arena of options, the college's menu prompts you to general information, degrees and admissions, student information, and alumni relations.

Obviously, as you'd expect, degree programs are introduced and thoroughly explained. From maintenance degrees to avionics, each area is summarized with no-nonsense prose. After researching the college's program options, test your aeronautics trivia skills with the monthly Air Challenge quiz. My quiz was 10 questions, and I'm happy to say that I passed easily—after skipping to the answers.

Free or Fee: Free.

Mountain Flying

http://www.mountainflying.com
e-mail: sparky@mountainflying.com

RATING

BRIEFING:

Generous assortment of mountain flying tips, complete with handy illustrations.

There's just no getting around them. Our world contains mountainous regions. And even plains-prone aviators should understand the important distinctions that peaks provide. Based on information from Sparky Imeson's book, *Mountain Flying Bible,* Mountain Flying's online resources dip into specific tips and teachings.

Almost an online textbook for mountain flying, Imeson's site speaks to basics as well as the advanced techniques of mountain flying. The chapters conveniently available from the left margin include "Must Know Info 1, 2 & 3," "Mountology," "Mother Nature's Tricks," "Book Reviews," "Scud Running," "Adverse Yaw," "About Stalls," "Emergency Landing," "Night Flying," "Spot Method," and "Oxygen." As you can plainly see, there are numerous topics here and lots of online reading. The writing is superb, typo-free, and augmented with an occasional illustration or table.

In my opinion, the best way to get the most from Mountain Flying is to go straight to the site map. Here you'll have a list of chapters and nice descriptions of each section's contents. Learn about stuff like mountain meteorology, density altitudes, leaning the mixture, downdrafts, updrafts/downdrafts, terrain modification, and more. What an excellent important resource.

Free or Fee: Free.

UND Aerospace–University of North Dakota

http://www.aero.und.edu

e-mail: online form

BRIEFING:

UND's Aerospace Sciences Aviation Department showcases its educational wares online.

The Aviation Department counts about 1200 majors in seven different programs. Students come from all 50 states and nearly 20 foreign countries. Good, now that I have your attention, click into UND's Aviation Department online composite for a well-described peek into why the school is so revered in aviation instruction.

One of the leaders in aviation-related training, the John D. Odegard School of Aerospace Sciences Aviation Department is a mouthful to say but is respected when said. Widely known as one of the major players, UND's Aviation Department promotes itself with expected grace online as well. Layout, presentation, text, and navigational components all mesh together perfectly with a harmonious hum. Just skim the seven-button menu to begin. Information relating to prospective students, current students, faculty and staff, alumni, visitor's news, and contacting the university make up your initial investigative options.

Then get a bit more enlightened as to the degree programs of air transport, commercial aviation, flight education, air traffic control, aviation systems management, and more. Still more details can be had in the way of classes and curriculum for each program.

Free or Fee: Free.

American Flyers

http://www.americanflyers.net
e-mail: info@amercianflyers.net

BRIEFING:
Handsome training and supplies presentation courtesy of American Flyers.

Emblazoned with its flying eagle logo, American Flyers soars onto the cyber-scene with expected grace, style, and worthy resources. Self-proclaimed "the world's finest pilot training since 1939," the American Flyers site expands beyond school information and instruction into an arena of pilot necessities as well as niceties. The flawless interface gives you a hint as to what lies beyond the startup screen.

High-flying page design lifts American Flyers into the elite group of Web professionals who know what they're doing. Use the fancy menu or a no-nonsense index to begin your journey. The Pilot Shop features headsets, flightbags, supplies, and more. Pilot Resources steers you into weather, forms, FAA documents, and cool aviation links. A Chart Room makes it too easy to avoid flying around with expired charts. Actual school information (domestic and international) is also available for those so inclined.

Then have some real fun with a free online ground school test prep. If you're rusty, sign up for the full-blown online ground school course (fee-related). It's handy 24-hour instruction for a variety of levels: private, instrument, commercial, ATP, and more.

Free or Fee:
Free. Online instruction series is fee-related with a few free samples.

The Academy

http://www.theacademy.net

e-mail: programs@theacademy.net

RATING

BRIEFING:

Class is in session on The Academy's professionally polished program presentation online.

With vistas of tropical Lakeland, Florida, in mind The Academy is one of the world's most unique vocational colleges offering three distinctively different programs. Your virtual Web tour takes you into The Academy's programs for the professional pilot, aviation maintenance technician, and culinary arts gourmet chef. While the latter may not be to your taste, the former programs do offer a tempting array of flying fare.

A helpful guide to those searching to fulfill career dreams, The Academy's site uses subtle, easy-to-read page layouts and steers clear of time-wasting graphics wizardry. Programs are carefully summarized, complete with snapshot photos of training in progress. Specifically, the Flight Program discusses occupations for graduates, certificates, ratings, and training conducted. Similarly, the Aviation Maintenance Technology Program area highlights general curriculum, airframe curriculum, powerplant curriculum, and course details.

Join The Academy online for thorough insight into its programs, credentials, and capabilities. Even an Admissions area answers your questions about acceptance, transfers, foreign students, and payment policy.

Free or Fee: Free.

FlightSafety International

http://www.flightsafety.com

e-mail: online form

BRIEFING:

Company information perfection from world-respected high-tech aviation trainers—it's what you would expect from these folks.

What better way to emphasize a professional approach to training than with a top-notch site. FlightSafety's expertise seems now to stretch into the cyber-arena. The company's History and Future introduction and site-wide content are written well, avoiding long-winded sales pitches. Content links to FlightSafety particulars give you up-front and current information. The introduction page serves up all your clicking options at a glance. Thereafter, you'll find the top frame searching to be satisfactory.

Content links include Course Schedules, Airline/Corporate New Hire Program, Maintenance Courses, Employment@flightsafety, Simulation Systems Division, Training Courses, and Military Training. Also flip through News & Events, Photo Gallery, Checklist, Challenger News, and FlightSafety Online (kind of an "e-learning" center).

There's a reason why corporations, airlines, the military, and government agencies rely on FlightSafety. Tap into this site and you'll get a feeling why.

Free or Fee: Free.

Aviation Communication

http://www.flightinfo.com

e-mail: instruct@flightinfo.com

BRIEFING:

Summed up succinctly as "serving the aviation community, as well as potential flyers who just want more information on aviation."

Aviation Communication refreshes the experienced and prepares beginners with an information-rich heads up. You'll fall smack into insightful tips and solid information from seasoned professionals. I even copied a few noteworthy tricks in Rules of Thumb, reminisced at the good advice in Checkrides, and refreshed myself with distances and separation in Airspace.

Don't expect online flying manuals or serious studying. This great source simply highlights and summarizes important topics. For example, Rules of Thumb combines helpful suggestions regarding descent, ground speed, wind computation, bank angle, true airspeed, horsepower, pressure altitude, temperature, climb, instruments, and airworthiness.

Although there's not enough room here for details, you'll need to trust me and spend time clicking into Learn to Fly, Build Flight Time, Aviation Chat, Aviation Medicals, Post Resumés, Airline Addresses, Classified Ads, Flight Schools, Instructing Hints, Aviation Quiz, and more.

It's aviator information euphoria. Enough said?

Free or Fee: Free.

Be A Pilot

http://www.beapilot.com
e-mail: online form

RATING

BRIEFING:

Dreamt of learning to fly? Fantasized about a well-organized intro-to-flight site? Touch down here for some encouraging reality.

Hey, stop dreaming and start browsing Be A Pilot's online taxiway to flight. Along with words of encouragement, you'll find visual page wizardry made up of handsome graphics that are quick to load. Omnipresent top and left-margin menus give you the guidance of a slick GPS. And plenty of efficient page layouts give you the "white space" to stay focused.

At the site's heart is an easy-to-use flight school search tool sortable by state. The surprisingly complete flight school list gives you many schooling options in your area, including brief descriptions and contact information. When you've narrowed your options and chosen a school, be sure to take advantage of the $35 introductory coupon. Just fill out the registration form (fairly lengthy) and print out the introductory flight coupon.

Other topics to call on include Welcome to Flying (nice, encouraging introduction), Aviation Links (companies involved with and endorsing Be A Pilot), Flight Schools Information Center, and the helpful Learning to Fly. This section gives you a quick but accurate peek into safety, steps to getting your license, costs, and more.

Free or Fee: Free, but make sure to fill out the survey and receive a $35 intro-to-flight coupon.

Applied Aerodynamics: A Digital Textbook

http://www.desktopaero.com/appliedaero/appliedaero.html

e-mail: info@desktopaero.com

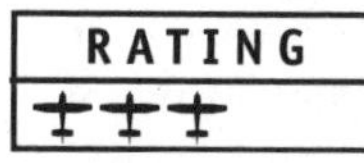

BRIEFING:

A scholarly online lecture that gives aerodynamic buffs a lift.

Kudos to Professor Kroo and the rest at Stanford University. This digital textbook arouses the awe of aerodynamics for aviators worldwide. Intended to supplement a more conventional aerodynamics textbook (the printed variety), the online version of Applied Aerodynamics peeks into some winged wonders via the Web. It's albeit a bit old. But the information contained will always be fresh.

Similar to a textbook, this digital resource provides thoughtful site organization with right-margin topics, a detailed table of contents, instructions, and an index. Although graphic design gurus weren't called on to send viewers into a GIF frenzy, the pages are simple and clean. Taking the place of visual perfection, interactive attributes within the text take the form of analysis routines that were built directly into the notes. (See, for example, the streamline calculations, airfoils, wing analysis, and canards.) You'll also stumble across cool charts and various depictions throughout.

Intrigued as to topics covered? Here's a taste: fluid fundamentals, airfoils, 3D potential flow, compressibility in 3D, wing design, and configuration aerodynamics. Sound like pocket protector stuff? Trust me, and jump into the slipstream—it's drag-free.

Free or Fee: Free.

SimuFlite Training International

http://www.simuflite.com
e-mail: info@simuflite.com

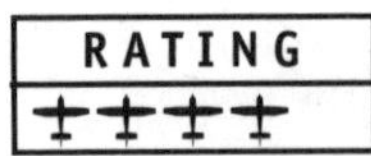

BRIEFING:

Advanced center for professional simulator training checks in with its advanced promo site.

I'll readily admit the reason I visited SimuFlite's site initially didn't include simulator training. I clicked in for the link to live conversations between air traffic controllers and pilots in the Dallas–Fort Worth (DFW) area. Of course, now there are many ways to get to the DFW live link, but I must say taxiing through SimuFlite's site opened my eyes to its quality offerings.

For those interested in simulator-based training, SimuFlite straps you into its cyber-stopover for a well-designed intro. Clicking into any of the following exemplifies the site's layout and organizational skills: About SimuFlite, Business Aviation and Training, Press Room, and Careers.

Even if you're not SimuFlite-bound, do visit the weekly Ground Chatter (in the Aviation Trivia area) for a collection of thoughts, sayings, and trivia. Depending on your capacity for nonessential entertainment, it might just be worth a bookmark.

Free or Fee: Free.

Neil Krey's Flight Deck

http://www.crm-devel.org/neilkrey

e-mail: neilkrey@aol.com

RATING

BRIEFING:

Welcome aboard the Flight Deck for an educated view on training and learning.

Whether you're relying on the instructor in the right seat for guidance or an airline pilot to get you home, experience factors into the background of most long-time professionals. Captain Neil Krey of Neil Krey's Flight Deck site has quite a background. This man, who is in the business of training and education programs, is an accomplished commercial pilot with an eye toward many aviation research areas: crew resource management, training and learning, the future, and more. In the Flight Deck you'll find links to and information about important topics such as Web-based training, aviation safety reporting systems, and scenario-based planning.

There's an educated look into aviation's future, as well as many fascinating studies and papers published by Captain Krey. Make yourself comfy in the jumpseat—you're going to learn something here!

Free or Fee: Free.

FirstFlight

http://www.firstflight.com
e-mail: tjs@firstflight.com

RATING
✈✈

BRIEFING:

No plane. No stalls. No talking to the tower. No expense. It's the cyber way to pilot a Cessna 152 for the first time.

Although a crafty approach to enticing new recruits, FirstFlight succeeds wildly in captivating potential flyers with some fun cyber-152 instruction. Excellent content and organization make FirstFlight easy to use for net novices as well as aviation novices. Unlike a flight simulator, this site steps you through the private pilot certification process via a series of "flights." Although not intended as a substitute for actual instruction, the evolving list of "flights" (new ones are added monthly) is a great preview to the real thing.

Scan through the current private pilot requirements. Read the private pilot syllabus. Examine checklists. Take an online preflight. And strap into the left seat. Following the script in each "flight," you'll become familiar with your cyber-Cessna: taxi, communicate via radio, take off, and land! You'll encounter embedded checklists throughout each "flight"—they're excellent references. Much of the information is basic and well suited to a flying introduction.

Congratulations to this innovative, educational, and interesting site. (How many sites can claim all these wonderful adjectives?) This guy's good—where do I sign?

Free or Fee: Free.

Embry-Riddle Aeronautical University

http://www.embryriddle.edu

e-mail: online form

BRIEFING:

A perfect demonstration of how smart folks (well, they are aviation educators) can grab the technological reigns and capitalize on appropriate Web applications.

From the minute you type its address, the Embry-Riddle page exceeds your expectations—even from the world's largest aeronautical university. This online equivalent of a university brochure wreaks of outstanding aesthetics while satisfying content-hungry surfers. Page navigation is simple, with well-designed omnipresent menus and clickable text.

Main topics are clearly labeled as Future Students, Current Students, Faculty & Staff, and of course, links to all its campuses (Daytona Beach, FL, Prescott, AZ, and extended campuses). Its content, quite simply, overflows the cup of perfection. From university information to research links to aviation links, you could easily spend days here. Education seekers will find admissions stuff, financial aid information, and a cool clickable campus map tour. Students (and anyone else) can tap into Career Info, the Avion Online (university-sponsored "zine"), and Associations. There's general campus news, library information, and more research areas. You'll even have at your fingertips faculty/administration information (including phone numbers), colleague information, and more.

In person or online, this university offers quite an education.

Free or Fee: Free.

Bookmarkable Listings

GG-Pilot
http://www.gg-pilot.com
e-mail: online form
Directory of America's top flight schools.

Aerofiles
http://www.aerofiles.com
e-mail: aero@aerofiles.com
Free online reference material for aviation historians, researchers, writers, and scholars.

Aeroflight
http://www.aeroflight.co.uk
e-mail: john@aeroflight.co.uk
Find detailed profiles of lesser known aircraft types, information on NATO and nonaligned European air forces, and more.

Aviation Aspirations
http://www.avasp.com
e-mail: online form
Worldwide depository of flight training recommendations.

AOPA Flight Training Magazine
http://www.aopaflighttraining.com
e-mail: online form
Thorough introduction to flight training from the world-renowned AOPA.

Flying-Explained.co.uk
http://www.flying-explained.co.uk
e-mail: ian@flying-explained.co.uk
Large database of U.K. flight schools and pleasure flight centers in the United Kingdom.

Wright Flyers Aviation
http://www.wrightflyers.com
e-mail: online form
Well-designed Web site detailing this flight school in San Antonio, Texas.

Joint Specialized Undergraduate Pilot Training
http://www.militaryaviator.com
e-mail: t@tung.org
Unofficial FAQs regarding the Joint Specialized Undergraduate Pilot Training—the U.S. Air Force program for producing professional military pilots.

American Academy of Aeronautics
http://www.americanacademy.net
info@americanacademy.net
California-based career flight school for aspiring professional pilots.

Western Michigan University College of Aviation
http://www.aviation.wmich.edu
e-mail: webmaster@aviation.wmich.edu
Huge online resource peeks into the offerings at Western Michigan.

Gleim Aviation Exam Preparation Materials
http://www.gleim.com
e-mail: webmaster@gleim.com
Aviation study/review books and software for professional exams.

King Schools
http://www.kingschools.com
e-mail: info@kingschools.com
Thorough online resource from King Schools—a world-renowned source for pilot learning and safety products.

Aviation Online Magazines and News

Flight International Magazine

http://www.flightinternational.com

e-mail: flight.international@rbi.co.uk

RATING
✈✈

Briefing:

Visually lackluster, this online aerospace news magazine dishes up dazzling insight and news.

Sort of like stumbling on some unwanted cows on the taxiway, two pop-up Web sites greeted me abruptly while logging into Flight International Magazine. I closed them quickly and moved into the meat of the magazine. Even if you run into a few pop-up sites, you'll find the hidden Flight International Magazine worth your precious time. Why? Simply because it's the online cousin to the world's number one weekly aerospace news magazine. And many of the printed magazine's current issue articles are neatly presented online.

Yes, maybe the flashiness you've come to expect with online "zines" is curiously lacking. Actually, the simple layout spills over into the ho-hum variety. But, hey, you're here for aerospace news. And aerospace news is what you'll get. Move into the Current Issue section and read some feature stories, as well as other weekly news articles sorted by categories (air transport, defense, business, general aviation, space flight, in-flight entertainment, and show report). The upside to Flight International Magazine is simply excellent worldwide coverage of aeronews. The well-written articles speak for themselves and alone deserve this award-winning mention.

Free or Fee: Free.

PoweredParaglider.com

http://www.poweredparaglider.com

e-mail: info@poweredparaglider.com

RATING
✈✈

BRIEFING:

Powered paragliding site packs a powerful punch with its premier online publication.

It's pretty amazing when you run across an online world you never knew existed. Especially if you scan hundreds of aviation sites daily. Well, not only does the powered paraglider world exist online, but it soars effortlessly with its powerful Web-based voice: Powered Paraglider. Touted to be "the #1 powered paragliding and paramotoring site on planet Earth," this expertly crafted compilation of everything paragliding serves its brethren well.

New to the paragliding world? Yeah, me too. Although I think I'll stick to my fuselage-encased flying machine, it's still fun to peek into the flying fancies of others. Custom tailored for beginners and advanced paragliders alike, you'll find topics like Getting Started in Powered Paragliding (PPG), Terminology, Books, The Sport, Photo Gallery, Movie Gallery, FAQs, Equipment, Technical Info, Training, Safety, Flying Locations, Clubs, and more.

Well designed, PoweredParaglider.com demonstrates how any information-driven Web site should be built. The introductory page has pretty much all your options at a glance down the left-margin menu. A Most Popular Pages pull-down menu steers you into commonly accessed pages. An informative Updates section sits atop everything, announcing site changes. And the balance of the introductory page details the different sections of the site with inline links taking you by the hand. Take heed Web developers, PoweredParaglider.com will show you the way.

Free or Fee: Free.

AviationZone.net

http://www.aviationzone.net

e-mail: info@avsupport.com

RATING

✈✈

BRIEFING:

Get *free* direct access to wholesalers, distributors, and manufacturers of aircraft parts, products, and services.

Do you crave airline and corporate aviation information, technical or otherwise? I may have unearthed a gold mine for you. Although it's fee-based (remember, you get what you pay for), AviationZone.net serves the airline industry and corporate aviation world with up-to-date and extensive news, portal, technical library, and e-commerce. Four insightful sections make AviationZone.net a hands-down award winner: News Desk, Industry Resources, Technical Library, and Buy & Sell.

Specifically, the News Desk serves up continually updated corporate information from PR Newswire and BusinessWire, with links into AviationNow.com. Leading aviation magazines such as *Airline Fleet Management, Regional Airline News,* and *Aircraft Technology Magazine* post current articles in this section too, adding to the many reasons why a bookmark is necessary. The Industry Resources section supplies a desperately needed industry-specific gateway to all things aviation. You'll find direct links to employee contacts and product/service offerings. The Technical Library (cream of the topic crop, in my estimation) stores a dynamic collection of technical documentation for airlines, manufacturers, and repair stations. Lastly, Buy & Sell promotes itself as being that quintessential online warehouse allowing links and access to third-party airline parts organizations, requests for proposals, and lists of products and services for sale.

Free or Fee:
Fee-based membership.

Wings Over Kansas

http://www.wingsoverkansas.com
e-mail: chancecom1@aol.com

RATING

BRIEFING:

Aviation news and historical perspective of interest for anyone who considers himself or herself an aviation enthusiast.

Sure, Kansas is a hotbed for aviation news. Heck, it's the "air capital" after all. So it probably goes without saying that Kansas aviation news can be had at Wings Over Kansas. However, for everyone who's a non-Kansas resident, you'll still want to add this bookmark. Why? Because there's so much more than just Kansas aviation news. In fact, the site is packed with interesting topics for all. Take, for example, the fantastic feature Aviation Profiles. This historical journal focuses on aviation heroes and notables with superb writing and photos. Current News, Learn to Fly, and Education are all topics you also should delve into on visiting Wings Over Kansas.

Oh, and Coming Attractions (as of review time) spotlights the treasure trove of goodies on their way to your clicking fingertips. Watch for information on flight school training for would-be pilots, a close examination of the general aviation community, a peek into the rapidly expanding corporate aviation market, the latest developments in commercial air transport, updates as to commercial and government air and space programs, and flying facts from the historical, modern, recreational, military, and commercial aviation perspectives.

Free or Fee: Free.

AvStop Magazine Online

http://www.avstop.com

e-mail: avstop@avstop.com

RATING
✈✈✈

Except for some typos, suspect site design, and annoying "Back" button navigation, AvStop Magazine remains a strong contender among the aviation "zines" of the online world. You probably know why already. Content.

BRIEFING:

AvStop will cause you to hold short of other Web wonders and browse its vast collection of articles.

Unless a drastic redesign occurs before or during your reading of this review, you'll also encounter a marginally presented collection of article links and photos. However, begin moving through the actual stories, facts, commentary, and articles, and you'll see the light too. Read about aviation news from the 1900s to the present, world aviation history, legal research tools for pilots and aviation enthusiasts, and general aviation stories. Specific site categories include Aviation News by Year, Aviation History World Wide, Legal Stories/Research Tools, Aviation Stories of Interest, Medical Stories of Interest, Books You Can Read Online (*Pilot's Handbook of Aeronautical Knowledge, Instrument Flying Handbook, Air Traffic Control Handbook,* and more), Practice the Airmen Knowledge Test Online, Technical Stories of Interest, and more.

Of course, plenty of pictures are sprinkled throughout AvStop to lend a little levity and tell the story as only a good photo can. Other aviation Web site links are also prevalent, but handy summaries keep you well informed before clicking.

Free or Fee: Free.

ATC Help

http://www.atchelp.com

e-mail: atctraining@home.com

BRIEFING:

An unprecedented perspective on the world of the air traffic controller (ATC). All aviators must visit. If there was a way to make it mandatory, I would.

Free or Fee: Free.

Ah, the ever-present air traffic controller. The terse, often busy voice behind the glass is human too, you know. Sometimes getting a bad wrap for snootiness or confusion, the ATC is, in fact, your best friend in the sky. Period.

One hundred percent helpful and always on your side, ATC Help devotes itself to educating the general public, pilots, and controllers about the ways and wonders of those in the tall tower. The real beauty of ATC Help lies not only in bookmarkable information but also in its grasp of perfect online delivery. Menus, instruction, headings, the What's New section, the monthly newsletter, and hints continuously guide you through the site—stuff that should be second nature for those who deal in navigation.

When you've familiarized yourself with ATC Help's tools and clickable features, dive right into some of the better site areas. Take a virtual tour inside the cockpit of a Boeing 777. Listen to any number of live ATC feeds (one of the largest collections I've seen on the Web). Visit ATC Jobs, ATC Forum, and ATC Articles to stay above the cloud deck on industry information. Read excellent training insight. The list of excellence goes on and on, and that's why they invented the bookmark.

Absolutely captivating for anyone is Gate to Gate, a multimedia experience that introduces you to the air traffic management system: the people, tools, and work of air traffic control. Just "right click" on these stored QuickTime files and begin your journey into The Early Years, Local Control, Departure Control, TMA, Surface Radar, Virtual Tower, Enroute Center Part II, TRACON, and Tower. It's a must-view movie for anyone in aviation.

Helispot

http://www.helispot.com

e-mail: info@helispot.com

RATING

✈✈✈

BRIEFING:

For helicopter stuff, this is the Helispot.

If only true magazine and informational Web sites would take heed of Helispot, the online world would be a much more user-friendly place. So simple. So clean. So straightforward. It's Web site navigational utopia. Why am I so worked up? Well, the Helispot home page is nothing more than an uncluttered collection of intuitive topics ready to be clicked. That's it. That's the navigation. No long, drawn-out descriptions. No rambling commentary. In fact, there's nothing but a topic interface.

However, just when you thought the site navigation was the golden nugget, the content glistens even more. View Recent Photo Additions, List of Categories, Helicopter Types, Helicopter Owners, Photos by Location, List of Helicopter Photographers, Photos from Specific Events, Classifieds, Links, Mailing List, and more.

Did I mention the helicopter slide show? Just enter how fast you'd like the pictures to change (cool!) and watch the show. The photos are captivating. It seems that almost every type of whirlybird is represented.

Free or Fee: Free.

PilotWeb

http://www.pilotweb.co.uk
e-mail: pilotmagazine@compuserve.com

RATING
✈✈✈

BRIEFING:

Revamped online "zine" that mirrors its famous printed sister: *Pilot* magazine.

Revolving around the United Kingdom's general aviation scene for nearly 30 years, *Pilot* magazine's printed version has made its mark and proven its worth to aviation enthusiasts worldwide. Now, with its redesigned online presence in the form of PilotWeb, the rest of us who don't subscribe to *Pilot* magazine can just click in for a virtual flight bag of information.

Not a U.K. flyer? Well, don't turn the page just yet. There's something for everyone here: book reviews, conversion tables, aircraft marks, forums, photos, past *Pilot* articles on instruction, flight tests, touring articles, and yes, a great reference resource for aviation acronyms, abbreviations, and jargon. Just click on a searchable alphabetical letter and get a succinct aircraft definition. From accelerate-stop distance to zulu, it's excellent for beginners and others puzzled by aviation jargon.

Incorporating only the best of the Web's current layout and site navigation components, PilotWeb contains many upgrades from its earlier online version. The clean, icon-rich pages are built for speed and subtle style. Lots of white space, text menus, and descriptions; excellent site search; and consistent layout throughout make the "new" PilotWeb a pleasure to fly. Grab the yoke and take a spin.

Free or Fee: Free.

Aerosphere

http://www.aerosphere.com

e-mail: editor@aerosphere.com

RATING

✈✈

BRIEFING:

Refreshing focus on aerospace tidbits that other sites seem to skip.

I must be honest and say that I almost zoomed right past Aerosphere. Aesthetically, it rivaled some of my least favorite sites. Instinctively, however, I scrolled through for a quick peek and found an award-winning reason for presenting it to you. Content. Lots of content. Original content.

As I wandered through this dazzling spectacle of aerospace news and tidbits, I realized that its entrancing qualities were worth a visit and then some. Seemingly catering to aviation and space lovers, this online magazine offers up a meeting place and emporium, with original and classic articles for the aerospace enthusiast. From aerobatics to UFOs to zeppelins, you'll find something that intrigues you. For full magazine article viewing you'll need a subscription (free as of review time), but the minor downtime is a small price to pay.

Traditional stuff like flight planning, weather, art, and a photo gallery are also waiting in the wings. However, the articles are truly worth your time. Oh, and Aerosphere's Amusement Park was a nice change of pace, with flight tests, aviation humor, games, and trivia.

Free or Fee: Free, but you must subscribe (free also) for full magazine contents.

The Aero-News Network

http://www.aero-news.net

e-mail: editor@aero-news.net

RATING

✈✈✈

BRIEFING:

News that will send any sport aviator into a satisfying spin.

Blazing trails of honesty and forthright reporting, the folks behind the scenes of the formerly printed *U.S. Aviator* magazine boldly beckon you to their newest online endeavor: The Aero-News Network. In it you'll find general and sport aviation news. Lots of news. Current news. Honest news. Actually, you'll be pleasantly bombarded with news, if that's your thing. Excellent reporting lures you in, and daily updates keep you coming back (a perfect reason to bookmark).

Prepare yourself for a dazzling if dizzying layout, riddled with news, flashing things, ad banners, frames, clipart, colorful text, and text tickers. Don't be intimidated, however. Site navigation really is effortless. Just realize that lots of text link menus will take you everywhere, while a top-margin frame grounds you in reality. There's even a guided site tour, and news searching is pretty easy too.

Often verbose, The Aero-News Network does aviation news well. Future site plans as of review time also may include aviation consumer watch investigations, a letters to the editor section, a media assistance center, an aviator store, an avionics database, classifieds, events, a picture gallery, and more. So stay tuned and fly by often.

Free or Fee: Free, but subscribe (just leave your e-mail) for e-mailed aviation news summaries.

Just Planes

http://www.justplanes.com
e-mail: info@justplanes.com

RATING

✈✈

BRIEFING:

Airline news and miscellany with a world-wide spin.

Here's a twist: easy, color-coded section descriptions on the introductory page. Now this I like. Although I've never flown into Just Planes, not even a flyby, I found myself instantly comfortable with the many airline-oriented choices before me. At a glance you'll see Airline News, organized by "current news," "next news" (thoughtfully listing the date when the updated news would arrive), and "past news"; Videos (for sale); Fotos (lots of airline photos from around the world); Airlines & More (links, forum, and show news); and Just Planes.

Click into the current Airline News to get an up-to-date refresher on news shaping the world of your favorite commercial carriers. Broken down into parts, part one emphasizes aircraft orders, leases, and deliveries. Part two features route information (you know, flights added to selected cities, etc.). Part three concentrates on miscellaneous airline news (mergers, acquisitions, agreements, etc.). And part four reports accidents and incidents.

Free or Fee: Free.

Mooney Owners of America (IMOA/MOA)

http://www.mooneyowners.com

e-mail: admin@mooneyowners.com

RATING

✈✈

BRIEFING:

Mooney madness prevails here.

No, not everyone's a Mooney owner. But there are enough owners out there who will find this site a worthy gold mine of everything Mooney. Have patience as you fly headlong into a wild cloud deck of ads, membership solicitations, clipart, and fonts of every size, shape, style, and color. Sure, it's busy. Yes, it's a bit confusing. But my goodness Mooney lovers, you've found your utopia here.

Reach into the omnipresent left frame for options o'plenty. Currently clickable options include Members Benefits, Join MOA, Site Intro/Home Page, Pilot Magazine, MLS & M20 Auction, Fly-Guide Intro, "Hot Stuff" Index, Flight Safety Foundation, MOA's Emporium, Classified Ads, M20 Hot Links, M20 News, Forum & Reader Feedback, and more. On deck as of review time are several additional online member benefits and programs, such as the Price Guide and Mooney Parts Clearinghouse/Locator Service.

Free or Fee: Some stuff is free, but you really need to sign up (fee-oriented) for all the Mooney madness you can handle.

Fly-Guide

http://www.fly-guide.com

e-mail: admin@fly-guide.com

RATING
++

BRIEFING:

Use this simple database to get the skinny on places to fly, buy fuel, and stay overnight.

Brought to you by the same folks at Mooney Owners of America, Fly-Guide caters to all pilots, not just Mooney owners. In a nutshell, Fly-Guide serves up an instant-access database of places to fly, buy fuel, stay overnight, play golf, and more.

Be forewarned: The design's not pretty. But that's okay. You'll get to the destination information soon enough anyway. Fly-Guide's listings are state by state and include several popular "fly-to countries." Each U.S. state or foreign country is further broken down into the nearest airport or city. Then click into the data and fellow pilot reports for the airports or cities listed. Once you find your city, you'll also be able to limit the pilot report listings by category (View All, Golf, Fishing, Water Sports, Airport Services, Eateries, Fly-Inns, Get-a-Ways, and Crew Car).

A nice feature worth note is the Airport Data box attached to some pilot reports. This gets you into data and facts that Fly-Guide displays from a previously published database. Although it's handy information on such things as runways and airport features, it's not always current. So, as always, consult the appropriate reference materials for current and correct information for flight planning.

Free or Fee: Free.

AviatorSelect.com

http://www.aviatorselect.com
e-mail: online form

BRIEFING:

Clearly a Canadian information resource, this megasite nevertheless will appeal to all aviators on some level.

Almost too big and intimidating, AviatorSelect.com serves up Canadian aviation topics o'plenty. I use *almost* because just the sheer size of this online information monster *should* be intimidating. Thankfully, it's not. This dazzling megamagazine of everything Canadian aviation wraps itself in perfect presentation and well-thought-out organization.

Of particular interest to most visitors are the aviation forums, in which visitors correspond with one another using a unique messaging system. Current discussions relate to the aviation industry, jobs, news, and general well-being. Certainly, though, there's more to AviatorSelect.com than forums. Current News, Link Directory (huge!), Weather, AircraftSelect.com (an excellent sister site focusing on aircraft for sale), and Surveys should keep you occupied for a while. Oh, and if aviation employment interests you, by all means step into the Employment section.

Currently evolving at review time is The Flyer. Directed at Canadian aviation enthusiasts, the content is submitted by visitors and will include reviews of aircraft and pilot accessories, aviation stories, and experiences.

Free or Fee: Free.

Airlinepilots.com

http://www.airlinepilots.com

e-mail: webmaster@airlinepilots.com

RATING

✈✈✈

BRIEFING:

Articles of fact, opinion, and speculation get you thinking at Airlinepilots.com.

I'm no airline pilot, but I know a few. Whenever I talk to them, they seem to be relatively up to date on industry news. Not sure about their sources. Maybe magazines or maybe talk in the crew lounges keeps them atop the latest in airline happenings. Those out of the know may need a little help. And that's where Airlinepilots.com comes in.

Although I stumbled a bit on faulty links and articles that seemed to have no content, my overall experience at Airlinepilots.com was positive. So much so that I'd recommend Airlinepilots.com to anyone who refers to themselves as "crew." Lots of article and feature-story links dribble down the introductory page, ready to connect you with the full story or a first-hand news source.

A long list of regular menu items begins your journey, along with a few sponsor-type links. Dabble around in Airlinepilots.com with Weather, Flight Tracking, Viewpoint, Interview, Legal, Labor, Finance, Aviation Movies, Contracts, Safety, Message Board, Chat, Ask Marty, and more.

Free or Fee: Free.

PlaneBusiness

http://www.planebusiness.com
e-mail: pbadmin@planebusiness.com

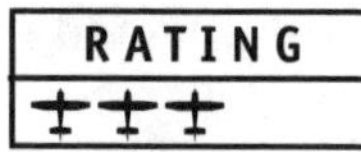

Briefing:

PlaneBusiness plays the role of airline industry publishers.

Anything but plain, PlaneBusiness is always on time with airline industry stats, daily business news, and financial tidbits. Better yet, it's pretty entertaining.

Written well, designed well, and updated daily, PlaneBusiness is your carrier correspondent full of news, insider talk, and dry wit. Presented in summary menu format, the features and articles load instantly with no need to get on the standby list. Just pick your linked topic and begin boarding: PlaneBusiness DailyBanter (the only daily financial wrapup of the airline industry), PlaneBusiness Banter (headlines and more from the subscriber-based weekly e-mail newsletter), PlanePerspectives, PlaneBusiness Message Boards (enjoy some airline industry buzz), and more.

Even if you're searching specifically for a favorite or not-so-favorite airline article, just type it in to reveal a list of all PlaneBusiness mentions. Or simply dig through the recycle bin for a mound of old airline-related stuff. It's all in there; just don't forget the rubber gloves.

Free or Fee: Everything's free site-wide, except for PlaneBusiness Banter (a weekly e-mail newsletter).

AeroSpaceNews' Leading Edge

http://www.aerospacenews.com

e-mail: online form

RATING

Briefing:

Mega multimedia medley of entertaining news, views, and movies.

I suppose when all is said and done, I probably would be considered a sucker for bells, whistles, and gizmos. But aren't most pilots? Anyway, if you're like me and enjoy a good multimedia online experience, I've uncovered a bookmarkable addition for you. AeroSpaceNews' Leading Edge "e-zine" incorporates just about every sense but smell.

Although not gracefully visual as Web pages go, the site's variety of aviation topics combines with cool sights and sounds to get you quickly involved. Begin with a quick look through your list of weekly news pages—General Aviation, Airline, Space, Military, Feature of the Week, This Week in History, Editorials and Commentary, Humor and Other Fun Stuff, and more.

A word of caution to those technically challenged: You'll need many plug-in applications to see and hear this site's good stuff. RealAudio, QuickTime VR software, RealVideo, and Java capabilities are simply a must. With the latest versions of either Netscape or Explorer, you should have no problem. But if you're lacking a specific plug-in, AeroSpaceNews provides handy links to free download areas.

Streaming NASA TV broadcast, the Virtual F/A-18 E/F Cockpit Tour, Audio of the TWA-800 Tapes, Airforce Radio News, and a Virtual Boeing B-777 Cockpit Tour are among my favorites.

Free or Fee: Free.

Aviation International News

http://www.ainonline.com

e-mail: editor@ainonline.com

BRIEFING:

An astounding array of features, current news, and pilot reports for the business aviation world.

Real reporting. In-depth insight. A bonanza of breaking news. Aviation International News (AIN) means business. With the precision and professionalism you'd find in today's corporate aviation world, AIN shines with "suit and tie" appearance and something between the ears too.

Matching its printed sibling in style and wit, AIN online displays its mastery in Web design and organization. The pick lists are clear. Articles are easily intelligible. And the Top News list spans the entire introductory page with summary links. Photos are few, articles are written well, and unobtrusive menus are everywhere.

Content-wise, AIN is just as enriching. Only one click away, the top stories touch on such topics as flight testing of new business jet aircraft, fractional ownership news, FAA & NASA studies, big-name FBOs, buying preowned aircraft, FAR commentaries, technology breakthroughs, and more.

As an added bonus, pilot reports pick apart the latest in business aircraft. Real pilots. Real analysis. They're just the sorts of reports that should be of interest if your company's shopping for some wings.

Free or Fee: Free.

AviationNow

http://www.aviationnow.com

e-mail: feedback@aviationnow.com

BRIEFING:

An expectedly satisfying variety of aviation news, resources, and stuff to buy.

Okay, let's pause and take a quick inventory of the Web's aviation "zines." Lots of monthlies. A fair amount of quarterlies. An unlimited number of those "randomly updated." And now it's "Now." Religiously updated continuously, AviationNow is everything a well-crafted Web "zine" ought to be.

Beauty and brains. Quick-link menu of resources. New featured stories teased on startup. All in-depth articles, well-written from the name you know: *Aviation Week & Space Technology*. However, here's the twist. You'll be invited to delve into a smattering of newsworthy tidbits from its major sister publications too: *Aviation Daily, Aerospace Daily, Business & Commercial Aviation, Airports,* and more. AviationNow's main menu leads you skillfully into worthy information for military and commercial aviation, space, business aviation, maintenance/safety, and more.

Many linkable choices can take you in many directions. There are lots of resources and pointers to sibling publications (McGraw-Hill being the proud parent): *World Aviation Directory, A/C Flyer, Overhaul & Maintenance,* and a host of aviation newsletters, just to name a few. Other aviation pursuits line the menu too, such as careers, a photo gallery, a reference center, upcoming events, and lots of stuff to buy in the store.

Free or Fee:
Some free articles. Printed magazine is subscription-based.

The Southern Aviator

http://www.southern-aviator.com

e-mail: webmaster@southern-aviator.com

BRIEFING:

Punch in Southern Aviator on your bookmark GPS if you're a southern states' flyer.

Bursting at its southern-style seams, the Southern Aviator's online complement to its print cousin serves up southern hospitality. The official voice for southern states' aviators (and those transitioning through) speaks volumes with such a handy information-rich resource.

Take, for instance, the daily postings of area news and events. You'll get a nice summary first with an invitation for more. Example articles deal with such topics as "Southern Carolina Increasing the State Share for Some Important Airport Projects," "Kissimmee Plays Host to What Could be the Largest Gathering of P-51 Mustangs in Recent History," and "An Historical Perspective of the Civil Air Patrol Wing Commander of Louisiana." Well, you get the idea. Topics vary, but southern-related news is always at hand.

Once you've read the latest, browse the left-margin menu of essentials. Serious resources, product presentations, and light-hearted entertainment find their way onto your screen. And Southern Aviator's sections steer you into the best of TSA, Calendar, Flying Places, Hangar Talk, Classifieds, FBO Showcase, Hotlist (links), Great Circle Distance Calculator, and more.

Are you a southern aviator? Hey, "smart birds fly south" by pointing their browser here.

Free or Fee: Free to browse, but there are many things to buy.

AirDisaster.com

http://www.airdisaster.com
e-mail: online form

Briefing:
AirDisaster is sometimes gruesome, sometimes enlightening, but always disturbing.

AirDisaster offers a well-organized forum for air safety and observations into preventable situations. The bulk of the content, however, delves into the disturbing pool of airliner accidents. And it is disturbing. Certainly photos are everywhere. Small and large. Graphic and innocuous. The continuously updated Statistics section compares accident rates by aircraft.

Relatively well organized and tasteful in presentation, AirDisaster's purpose isn't to glorify the gruesome. Rather, its offerings combine a series of articles, reports, and statistics identifying the nature of airliner accidents. Clickable topics quickly move you into Live Chat, Air Safety Forum, Accident Database, Voice Recorders, Photo Galleries, Accident Movies, Special Reports, Eyewitness Reports, and more.

Note: The subject matter lends itself to disturbing depictions of airliner accidents in the form of photos and video.

Free or Fee: Free.

Plane & Pilot Magazine

http://www.planeandpilotmag.com

e-mail: editors@planeandpilotmag.com

BRIEFING:

Popular Plane & Pilot roars onto the Web's cyberscene with pilot resources and a few articles.

The popular magazine for active piston-engine pilots, *Plane & Pilot* teases a bit with its online offerings. The presentation and left-margin menu are certainly capable enough for the interested surfer, but content serves largely to whet the appetite for the printed publication. Get a handy list of summarized feature stories in this month's and past issues. Generally, one feature story is offered in its entirety.

Plane & Pilot magazine has been one of my favorites for years. So it stands to reason that your reviewer would include its site as an award winner if it brought its standards of excellence to the Web. I think you'll agree that its online sister is a winner, not for savvy self-promotional offerings, but for its schools directory and monthly feature article. Obviously trying to be thorough, Plane & Pilot's schools directory links up to all the majors and many minors in the category of aviation training. Relatively comprehensive in scope, you'll be pleasantly surprised with such a quality collection.

Even if you're not in search of training, Plane & Pilot keeps your attention online too with full-blown topical features monthly. Scan the back issues if you crave a bit more.

Free or Fee: Free.

Air Safety Online

http://www.airsafetyonline.com
e-mail: online form

BRIEFING:

The grim reality of airline disasters hits home and prompts this thorough call for safety.

Publishing "not just the facts" on airline safety and fatal disaster, Air Safety Online serves up a bit of editorial comment, too. Yes, you'll find horror and heroes, tragedy and transcripts. But Air Safety Online still seeks to encourage safety in the skies. Here's some site perspective: "You would have to take a random flight every day on a commercial aircraft for almost 30,000,000 years to be insured of being in a fatal plane crash."

The fact is, airline disasters do happen. Maybe not to you or someone you know. Nevertheless, we are all still affected by the news. In Air Safety Online, the light shines bright on past tragedies—ones you may have forgotten. Swissair 111, Delta Flight 1141, TWA Flight 800, and others come to life in words, pictures, and voice recorders. Site options include a message board, online reports, photos, videos, statistics, voice recorders, links, and more.

Sprinkled into the mix of history, pictorials, facts, and statistics is the author's editorial thoughts. Almost surely provoking your ire or agreement, the opinions are candid and earnest.

Free or Fee: Free.

AirConnex

http://www.airconnex.com

e-mail: online form

RATING

BRIEFING:

Award-winning site for airline news and air travel links.

At first AirConnex appears a bit simplistic: nice design, fast loading, and just a few summarized site categories. Certainly scud-running the cyberskies of gaudiness, AirConnex has chosen a new route. A less complex and flashy route in its quest to "answer as many of your air travel needs as possible." Where's the huge out-of-scale plane pictures and endless dead-link directories found at most information sites? Not here. Climb aboard yourself and see what I mean.

Mostly text description greets you on startup, which foils most time wasters. Each category is revealed before clicking with a more than adequate peek into content. Probably harboring the most value is the wonderfully independent Air Bulletin. This weekly global airline newsletter keeps you current on commercial carrier news and events, complete with photos and side links. The Aviation Bookstore serves up a nice interface for book ordering (via Barnes and Noble) with such categories as accidents, aircraft, history, pictorials, travel, airlines, and more. Or if you're just searching for your favorite airport, tap into an easy-to-use compilation of airport Web site links.

Free or Fee: Free. Leave your e-mail address for future updates.

CyberAir Airpark

http://www.cyberair.com

e-mail: webmaster@cyberair.com

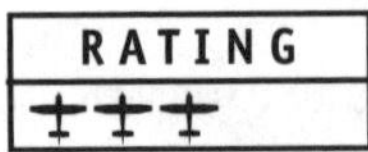

BRIEFING:

Touch down at the Airpark for tantalizing tidbits, prizes, entertainment, and handy safety stuff.

Once you land at CyberAir Airpark, you've got a myriad of stops right off the taxiway. Click on an Airpark map or select from the site index for Control Tower and FAA Info, The Museum, the RealAudio Area (have your audio player geared up), the RealVideo area, CyberAir FBO, Links, and more.

Although the preceding locations certainly warrant a visit, you may want to skip directly to Hot Stuff selections, including Last Week's Aviation History, Free BFR at CyberAir Airpark, Aviation Safety Pages, and a fascinating RealAudio seminar show (at review time, I listened to a 59-minute presentation about the Boeing 777).

Page navigation skims the surface of simple, with clickable map or index options. Nifty additions include a scrolling information banner and a link to Chicago Approach Control—live!

Free or Fee: Free.

AeroWorldNet

http://www.aeroworldnet.com
e-mail: online form

BRIEFING:

International aerospace news that's updated daily and written well—the perfect ingredients for a bookmark.

The global aerospace perspective—you'll find it online at AeroWorldNet. Current and updated daily, this international aviation news source gives you efficiency and excellent insight. Topics include lead stories, weekly headlines, news, aerospace events, and more. Great short summaries capture the essence of most articles before you click into the full story.

Mostly text-based with some banner ads and logos floating around, the pages are quick to load and easy to navigate. Left-margin links take you to additional subjects, such as Aerospace Jobs, People and Places, Industry Literature, Industry Products, Aerospace Events, Industry Message Board, Aerospace Companies, Industry Associations, and Membership (sponsorship) in AeroWorldNet.

With not many places to turn for solid aerospace news online, you will be rescued when you land on AeroWorldNet's informational oasis. Lacking are a plethora of typos, giant pictures, and old news. It's a world of positive difference compared with your other Web options.

Free or Fee: Free memberships—sponsorships are available.

Planet Aviation

http://www.planetaviation.com

e-mail: tower@planetaviation.com

BRIEFING:

Web radio show just for us aviation types.

A radio show on aviation? Yes. But better still, you don't need a radio, and you can tune in live—worldwide! Self-dubbed as "where the best of aviation hang out every single week," Aviation Weekly rides the radio waves of Web innovation and brings aviation enthusiasts this unique approach to industry news.

Traveling at 186,000 miles per second, Aviation Weekly's aeronautical "audiology" comes together into entertaining shows. Left-button topics move you into your audio preferences of archived and live radio shows. Lots of show description before you link up make show selection easy. Specifically, the huge archived show list highlights topics covered. Just click the speaker icon to load the show.

Crank up your 56-kbps (or hopefully faster) modem, and point your browser here to hear.

Free or Fee: Free.

FlightWeb

http://www.flightweb.com

e-mail: rparrish@flightweb.com

RATING

✈✈

BRIEFING:

An air medical must-see. Check in for current, thorough industry information.

Faster than you can say "E.R.," FlightWeb bursts onto your screen and serves up no-nonsense air medical resources. Homebuilders have their own sites. Aerobatics buffs have theirs. And now air medical professionals have an online forum to gear up for air medical transport readiness.

This dedicated breed of specialists comes together with their own well-organized cyber-support staff. Not limited to just pilots, FlightWeb dips into content designed for worldwide air medical folks who are flight medics, nurses, medical doctors, communications specialists, dispatchers, and more.

Your blue-button menu links you up with a wide range of worthwhile reasons to add a bookmark. Industry professionals will want to scan through Air Medical Web Pages—EMS, Medevac, Associations, Aviation, Flight Programs, Vendors, and others; Chat Room; Flightmed Mailing List Archives—a searchable grouping of posted messages; Resources—air medical clipart, mentors, medical protocols, legal issues, and various FAQs; White Pages—e-mail directory of air medical pros; and more.

Scrub up and join FlightWeb in this well-educated online operatory. The air medical doctor will see you now—24 hours a day.

Free or Fee: Free.

Aviation Safety Network

http://aviation-safety.net

e-mail: harro@aviation-safety.net

BRIEFING:

An award-winning yet dark view into the world of airliner disasters.

Morbid and sometimes somber, Aviation Safety Network may seem at first to cater only to the accident curious. Those who are drawn to the disaster scene with fascination will certainly get their fill of fatality statistics and gory details. If you reach a little further, though, you'll notice that this airliner accidents site employs descriptive accident data—the kind that may lend a hand with your own accident-prevention preparation.

Religiously updated, this site contains a wealth of airliner accident information found in features like The Database, Statistics, Accident Reports, CVR/FDR (Black Box Transcripts), Safety Issues, and Accident Specials. Lists of accidents by year reveal endless summary tables of data. Event details such as aircraft type, operator, and flight route combine with phase of accident, fatality counts, and remarks to paint a graphic accident picture. Links to additional articles, photos, and other reference sources round out your research options.

Have an unfulfilled curiosity about airliner disasters? New fatal airliner accidents are added within one or two days. Sign up to receive free e-mail "digests" to keep you apprised. Site update notification is also available for the asking.

Free or Fee: Free.

FlightLine OnLine

http://www.aafo.com

e-mail: submit@aafo.com

BRIEFING:

This "quasi-zine" (the company has steered away from the traditional magazine boundaries) serves up flying features and news.

When an aviation information site offers such an obvious dedication to current updates, I'm immediately interested. Combine continuous updates with fascinating feature stories, and I'm hooked. FlightLine has the "zine" ingredients most have come to expect, but the site's flair for variety is unique.

Sort through great air race information, gallery photos, feature articles, news, links, and flight simulator information. Read the latest news or topical features worldwide, or delve into some fun audio and video clips. Whatever your fancy, a constant text menu flies right seat with you—always ready with new topic coordinates. Expect to uncover the latest in global air show news, be bombarded with a great collection of photos in the gallery, and see pointers to some highly regarded links. Finally, this award-winning information gem rounds out your cyber-options with flight simulator news and tips and lots of discussion in the Hangar Talk message board.

Other innovations you may appreciate include the latest news from the American Forces Radio News and Department of Defense News Briefs.

Free or Fee: Free.

AVWeb

http://www.avweb.com

e-mail: editor@avweb.com

BRIEFING:

Free membership after completing a survey entitles you to competent aviation-related journalism.

A daily information resource. Yes, daily. One of the best, most competent online aviation publications around. Get your news each day, or browse in summary form with the e-mail edition—AV Flash—every Monday and Thursday. The managing group consists of seasoned professionals—writers, editors, and publishers with years of experience. You'll recognize names of well-known regular contributors—the best in aviation journalism.

Selections include NewsWire, ATIS, Safety, Airmanship, The System, Avionics, Places to Fly, Reviews, Shopping, Classifieds, Brainteasers, Weather, Net Sites, and more. Also dip into an organized database of FAA Aircraft, Airman and Mechanic Registries, the Medical Examiner and Repair Station Lists, U.S. Airman Directory, the FARs, and others.

You'll find a host of site-navigation features, including menus everywhere you go, clickable topic icons, and convenient descriptions. I've been an avid subscriber for many years now, and let's just say it's tops on my bookmark list.

Free or Fee: Free, but sign up for all the good stuff.

Aerospace Power Chronicles

http://www.airpower.maxwell.af.mil
e-mail: editor@cadre.maxwell.af.mil

BRIEFING:

Brought to you by the folks at *Airpower Journal,* Air Chronicles gives you online insight into today's modern Air Force.

Air University Press, publishers of *Airpower Journal,* have opened the door to this interactive Air Force information resource. It's about Air Force doctrines, strategy and policy, roles and missions, military reform, personnel training, and more. Jump in and submit an article, or read what others have to say.

Categories include Airpower Journal, Contributor's Corner, Airpower Journal International, Current Issues, and Book Reviews. My favorite, Contrails, gives you a mixed bag of aviation stuff, such as news, weather, tools, references, government sites, symposia, professional development, fiction, and professional journals.

Military authors, civilian scholars, and people like you and me create this heads-up, intelligent review of our modern Air Force.

Free or Fee: Free.

In Flight USA Online

http://www.inflightusa.com

e-mail: staff@inflightusa.com

BRIEFING:

Here's an in-flight magazine you'd actually want to take with you. Lucky for you this handy little "e-zine" is on the Web—no need to waste an airline ticket to get it!

Stow and lock your tray table. Return your seatback to its upright position. Give your flight attendant all that garbage you're holding. And most important, don't forget to snag that in-flight magazine! Okay, truth is, they're only good for a captive audience. However, this online magazine (also existing in print) piques your in-flight interest a little better.

Soaring with juicy, current aviation information (stemming from each monthly printed version), you'll tap into a variety of hot topics. Dive into news, editorials, features, and executive reports. Or get information on a variety of special topics such as flight instruction, homebuilts, air shows, new products, and more. During your online stopover, you'll also be invited to get side-tracked with a nice selection of stable links (lots of NASA stuff).

Oh, and if you just can't fly without the printed version, you can simply e-mail a subscription request. The company makes it so easy.

Free or Fee: Free for online viewing, fee for printed magazine subscription.

European Business Air News

http://www.bizjet.com/eban

e-mail: richard.evans@bizjet.com

BRIEFING:

A down-to-business air news site for European business aircraft owners—efficiently designed to foil time wasters.

Enough flying fun, let's get down to business. With European Business Air News you immediately get the impression that dedicated, responsible adults are pulling the strings on this site. It's mostly text, organized well, with a few photos (I'm a fan of bandwidth efficiency).

Specifically focusing on daily European business aviation news, the opening page has links to all the current events you can handle. Some newsworthy article teasers load initially by default (from most current to least current). Then, if the need arises, click on News in Brief topics or scan back issues with past articles.

After you've caught up on today's headlines, you'll find anything else news-related that suits your fancy with a streamlined search box. Just type in your topic, and away you go to good news.

Free or Fee: Free.

Journal of Air Transportation World Wide (JATWW)

http://www.unomaha.edu/~jatww

e-mail: journal@unomaha.edu

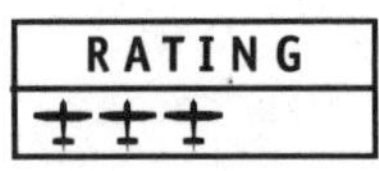

BRIEFING:

A scholarly online endeavor that will pique interest among enthusiasts.

Before sifting through this site, it may be appropriate to fix your bow tie, position the specs on your nose just right, and have aviation theory volumes at the ready. Before intimidation sets in, browse through the offerings—you *will* learn something. The journal's goal is to eventually become "the preeminent scholarly journal in the aeronautical aspects of transportation." Lofty? Yes, but the JATWW will offer an online sounding board for peer-reviewed articles in all areas of aviation and space transportation, research, policy, theory, practice, and issues. After article review, approved manuscripts will circulate via list server free to all subscribers.

At year end, bound volumes, including all accepted manuscripts, will be available for sale and library reference. If you're not interested in being a part of the free global distribution, you can still visit the site and review any articles of interest. Topics include aviation administration, management, economics, education, technology and science, aviation/aerospace psychology, human factors, safety, human resources, avionics, computing and simulation, airports and air traffic control, and many other broad categories.

Free or Fee: Free.

The Avion Online Newspaper

http://avion.db.erau.edu

e-mail: see site for appropriate e-mail

BRIEFING:

An online university "zine" that spreads its wings and intrigues more than just its fellow students.

No stranger to aviation news, the talented folks at Embry-Riddle Aeronautical University continue journalism excellence with this award-winning online newspaper. Mostly written for students, The Avion Online harbors a wealth of fascinating information for every flying enthusiast. You'll uncover the most current events in aeronautics. Find headline news makers on the front page, and enjoy a host of University stuff. Categories include Campus News, Metro News, Student Organizations, Space Technology, Data Technology, Diversions, Opinions, Sports, and Comics.

Organized well with ease of use in mind, clickable topics in the left-margin frame make navigation simple. An online search engine requires only a concept or keyword for subject look-ups.

Billed as "for students by students," The Avion Online gives us all an educated peek into aviation.

Free or Fee: Free.

Air & Space Smithsonian Magazine

http://www.airspacemag.com

e-mail: webmaster@airspacemag.com

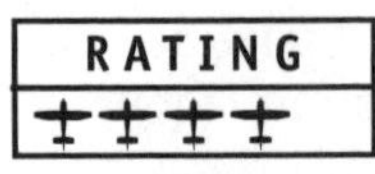

BRIEFING:

An online "zine" sibling to the printed version with its own distinct personality.

As your collection of unread aviation magazines continues to pile up (due to too much flying time), stroll on over to the keyboard and tap into this award-winning online magazine. Sure, you'll be encouraged to subscribe to the well-respected print version, but if you're interested, the online way is easy.

Just here to browse? Well, stay awhile—it's worth it. The Air & Space Web site gives you a peek into the current hard-copy issue and provides many additional online articles and features. For quick text-based navigation, the table of contents lists everything, including pages not referenced in the home page. Capitalizing on a history of journalistic excellence, Air & Space continues the tradition here with fantastic current events, interviews, reviews and previews, marketplace, associations, and more. You'll even find a nice selection of QuickTime movies and QuickTime virtual reality panoramas.

Free or Fee: Free (fee for magazine subscription).

WebAviation

http://www.webaviation.com

e-mail: webmaster@webaviation.com

RATING
✈✈

Briefing:

International cornucopia of aviation news, services, and miscellany.

French, English, Spanish, or Dutch. Pick your pleasure, and move into a nicely organized collection of thoughtful aviation topics.

Easy navigation is the rule at WebAviation. Main topics are manageable in number, and subtopics zero in on your specific preference. Choose from Services (headline news focused on aviation, specialized directories, events, classified ads, and mailing list), Flight Briefing (flight-planning information for licensed pilots), Learning, Shop (books only as of review time), and a few links in Other Sites.

That's it. Good, solid information, not a lot of fluff, wrapped in a easily navigable shell. I'll take a double order of that please.

Free or Fee: Free.

Jane's Information Group

http://www.janes.com

e-mail: janes.webmaster@janes.co.uk

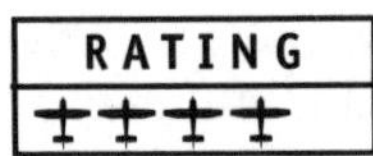

BRIEFING:

Almost reaching Web perfection overload, Jane's showers you with defense, aerospace, and transportation resources.

For those who may not know, *Jane's* refers to the legendary English pioneer John Frederick Thomas Jane, who began the widely regarded source for defense, aerospace, and transportation information. How appropriate that the world-renowned resources of Jane's have found their way onto the cyber-information highway. Jane's boldly celebrates more than 100 years of information solutions with its stunning example of Web design excellence.

Fascinating information and thorough descriptions are everywhere. Find out more about Jane's with Jane's Services (products, data service, consultancy, and news briefs), Aerospace (news, aerospace products, market review, contact points, word index, and links), and Magazine Sites (Defense Weekly, International Defense Review, Missiles and Rockets, Airport Review, and Transport Finance).

Don't misunderstand. There's a lot of publication selling going on here. However, from the people who brought you the bible of the aviation industry, *Jane's All the World's Aircraft,* you can expect to be more than satisfied with brilliant content.

Free or Fee: Free to view, but a myriad of informational products are just waiting to be ordered.

Bookmarkable Listings

The Controller
http://www.aircraft.com
e-mail: feedback@aircraft.com
Sneak peek into *Controller* magazine with for-sale classified and broker listings.

UK Airshows
http://www.airshows.co.uk
e-mail: paul@airshows.co.uk
Personal U.K. air show reviews, complete with future show information and past pictorials.

AeroCrafter
http://www.aerocrafter.org
e-mail: aerocrafter@eaa.org
Online complement to the printed magazine offers home-built aircraft information and sources.

United Space Alliance
http://www.unitedspacealliance.com
e-mail: communications@usahq.unitedspacealliance.com
Online space information center features space shuttle news, virtual facilities tours, and space-related resources.

National Championship Air Races
http://www.airrace.org
e-mail: online form
Official news and information site of the Reno Air Racing Association and the National Championship Air Races.

Ultralight News
http://www.ultralightnews.com
e-mail: ulnews@ultralightnews.com
Ultralight news madness runs amok.

GPS World Online
http://www.gpsworld.com
e-mail: editorial-gps@gpsworld.com
GPS stuff and then some gets pinpointed from Web to your screen.

General Aviation News & Flyer
http://www.ganflyer.com
e-mail: see site for appropriate e-mail
Online "zine" mirroring the printed one—no teasing here, though, there's plenty of articles and features.

Rotorcraft Page
http://www.rotorcraft.com
e-mail: norman@biztool.com
Here's enough information to keep your head and your gyrocopter spinning.

US Aviator
http://www.av8r.net
e-mail: none available
Current aviation news that won't waste your time.

SW Aviator Magazine
http://www.swaviator.com
e-mail: publisher@swaviator.com
Southwest-specific online magazine complements its printed twin nicely.

PlaneCrashInfo.com
http://www.planecrashinfo.com
e-mail: kebab@tstonramp.com
Comprehensive airliner crash database with facts, figures, photos, and links.

The Aviator Web Site
http://www.aviatorwebsite.com
e-mail: info@aviatorwebsite.com
Well-rounded resource for aviation employment, various directories, and information on learning to fly.

Armed Forces Journal International
http://www.afji.com
e-mail: afji@afji.com
Monthly international defense journal.

Business Air Today
http://www.businessair.com
e-mail: none available
Elite publication targeted toward buyers and sellers of corporate jets.

Flight Journal Magazine
http://www.flightjournal.com
e-mail: flightjournal@airage.com
Online magazine of rare aircraft, extraordinary people, and remarkable stories.

SPEEDNEWS
http://www.speednews.com
e-mail: aviation@speednews.com
Weekly newsletter for the commercial aviation industry designed for a "quick read."

Aviation Parts, Supplies, and Aircraft

The Pilot Shop

http://www.pilotshop-usa.com

e-mail: info@pilotshop-usa.com

RATING

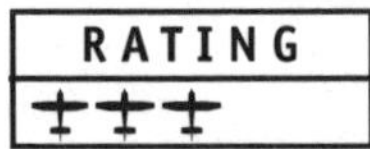

BRIEFING:

The Pilot Shop. A name so apropos, an online shop so inviting.

The last thing I want to do after a difficult, navigation-intensive flight is to wrestle around with my aviation shopping online. Frankly, I look for two main things: fast and easy. I prefer not to be vectored around airspace that isn't part of my flight plan. With The Pilot Shop, you get form and function, style and substance. Just scroll through the left-margin menu of aviation necessities. Lots of main categories represent almost anything you could possibly need: aircraft accessories, apparel, books, charts and maps, cockpit aids and accessories, flight bags, supplies for women aviators, gifts, GPS receivers, software, scanners, videos, and more.

The best part is, once you lock into something of interest, you merely click the category and your text menu of products appears in the main frame instantly. No pictures. No ads. Nothing to slow you down. Next, select your product from the menu and get a minipicture, nice description, and opportunity to order.

Physically, The Pilot Shop sits on the east side of the DuPage Airport in West Chicago, Illinois. Next time you're flying by, stop in. Once on the ground at DuPage, just say to ground control: "Taxi to Pilot Shop," and they'll give you a progressive taxi right to the front door. Now that's service.

Free or Fee: Free.

The Eastern Avionics Guide to Avionics

http://www.avionix.com

e-mail: staff@avionix.com

RATING

✈✈

BRIEFING:

Zero in on avionics information, tips, reviews, and products from the cockpit-savvy crew at Eastern Avionics.

Free or Fee: Free.

Making our world aloft easier and safer, avionics is a topic near and dear to us aviator types. Gizmos, gadgets, knobs, bells, and whistles get my heart rate up. I suppose it's the techno junkie in me. So imagine my drooling state as I climbed aboard the Eastern Avionics Guide to Avionics.

Since 1995, Eastern Avionics' Web site has attempted to provide up-to-date information on nearly everything in light general aviation avionics. And it shows. Maybe a little too much. My internal GPS fizzled a bit as I glided around the Web site relying only on dead reckoning. There are so many menus and options and links that even my old VOR became disabled. However, rest assured that you'll eventually find your avionics products, tips, information, reviews, and links here.

A few standouts became clear quickly in my site searching. The excellent Pilots Guide to Avionics helps to steer you in the right direction with essential information, tips, reviews, and important issues that all pilots should consider when shopping for avionics. Equally informative, but with a multimedia spin, are the many "roundtable" discussions found in the Avionics Radio studio. Have your RealAudio player on standby, and click into audio discussions about GPS systems and the IFR environment, tips on selecting the perfect headset, slide-in radio replacement tips, great buys in reconditioned avionics, and more.

All Aviation Gear

http://www.allaviationgear.com

e-mail: info@allaviationgear.com

BRIEFING:

Loads of supplies grace the pages of this well-designed aviation product wonderland.

Wow. You don't often come across such a combination of flash *and* speed during your online shopping adventures. Usually you give up something. Aesthetics for efficiency. Or vice versa. However, with All Aviation Gear you'll cruise at red-line speed with all-star functionality. Frame-driven pages divide your menus, searching, and content. Search by New Products, Hot Deals, or a Product Search Box. Excellent product presentation appears, complete with description, photos, and prices.

Part of what makes All Aviation Gear so appealing lies in its well-thought-out organization. For example, a top-frame menu takes you into specific areas of products based on your needs. Choose from Student Pilot, VFR, IFR Pilot, Commercial Pilot, and Hangar Gear. Featured products are appropriately tailored to your skill level or need (nice!).

When you're ready to buy, a better-than-average shopping cart system takes you by the hand. You can purchase if you're not a member. However, set up an account. The company will retain your information to make your second visit to All Aviation Gear a pleasant one.

Free or Fee: Free. But members receive special offers, newsletters, and more. So sign up. It's free.

Oshkosh Pilot Shop

http://www.oshkoshpilotshop.com

e-mail: webmaster@oshkoshpilotshop.com

RATING
✈✈✈

BRIEFING:

When shopping online, those who feel the need for speed should check into the Oshkosh Pilot Shop.

Oshkosh in Florida? Yeah, I was confused too at first. Your visions of 3000 show aircraft that make their way to Oshkosh, Wisconsin, don't really apply here. Oshkosh Pilot Shop is actually located in Miami, Florida. Compliments on the name, though. It got my attention. Actually, Oshkosh Pilot Shop is Miami's biggest and most complete pilot shop (its claim, not mine), located right across the street from Miami International Airport and offering all types of pilot supplies, accessories, and gifts at competitive prices.

It's an okay shopping site in many respects. You know, categories and products, descriptions and pictures. The shopping cart is always at the ready in case you've just got to have something. The speed of the site, however, *is* worth a more enthusiastic mention. Shopping is quick and easy with page loads that fall into the instantaneous variety. Product scanning and reviewing become pretty painless while scrolling through the Oshkosh Pilot Shop online.

Ready to cruise the isles? Click through their left-margin buttons of GPS, headsets, log books, flight bags, charts and maps, books, gifts and apparel, software, transceivers, current weather, new products, and computerized aircraft systems.

Free or Fee: Free.

AviaBid.com

http://www.aviabid.com
e-mail: admin@aviabid.com

RATING
+++

BRIEFING:
Online aviation auction sparkles with flawless function.

Launched back in September of 1999, AviaBid.com dazzles the online aviation world with what it deems to be the "Internet's first and only aviation auction community and marketplace for the aviation consumer and professional buyer and seller." Not sure about the *first* and *only* parts, but I do know that AviaBid.com spent some quality time developing a first-class design and navigation system for its auction. It's flawless. The auction part is solid. And help features are everywhere.

Hopefully by now you've at least heard of online auctions, if not used one. No? It's aviation buying and selling through private parties with AviaBid.com as the middleman. The seller picks a length of time for bidding. Bidding automatically terminates at the end of the auction. At the end of the auction, the sellers and winning bidders are notified via e-mail with all transaction details listed. There are other processes and auction details to know, but you get the idea.

Membership in the auction community is free, and the categories are many: aircraft, aircraft parts, books, pilot accessories, software, tools, and more. The downside? Well, at review time the auction items were few. However, by the time you read this, AviaBid.com undoubtedly will blossom as a bookmarkable gem.

Free or Fee:
Membership is required but free.

HistoricAviation

http://www.historicaviation.com

e-mail: customerservice@historicaviation.com

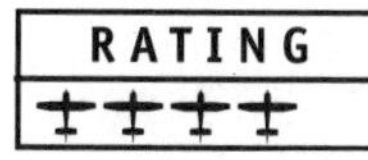

BRIEFING:

Humble to humongous, HistoricAviation supplies you with thousands of aviation goodies online.

From its humble mail-order beginning more than 20 years ago, HistoricAviation has spread its wings online with thousands of aviation items, including books, videos, art, models, kits, puzzles, and more. You'd think that the gigantic assortment of aviation goodies found at HistoricAviation is reason enough to visit. And you'd be absolutely correct. However, the online presentation deserves your attention, too. This perfect layout quietly keeps you organized and content while you navigate easily through the pages. Drop-down menus provide the product search based on product type (books, videos, art, models and kits, calendars, and more) as well as on category (World Wars I and II, other military, flying, and civil aircraft). And lots of feature products line the pages.

Carefully enclosed in a protected shell of a secure server, HistoricAviation peddles its wares with a well-designed shopping cart. It's relatively easy to buy, but the product pictures were a bit small at review time. Maybe that's why the site's pages are so quick to load. Overall, it's fun, fast, and friendly. Whether you're looking into buying a plastic model of the historic "Tin Goose" flown over the South Pole in a 1929 Antarctic expedition or you just want to pick up the video *How to Fly the B-24*, it's all here.

Free or Fee: Free.

AvBook.com

http://www.avbook.com
e-mail: none available

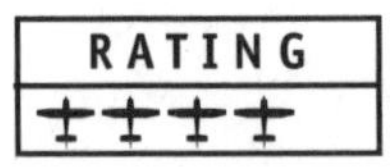

BRIEFING:

Its name tells the whole story. Aviation books, plain and simple.

One word comes to mind during my online visit to AvBook.com: *solid*. This nicely wrapped package of book-buying goodness covers all the basics, presents thoroughly, and sells securely.

The all-aviation shelves at AvBook.com stock all the top aviation titles. Yes, the company actually stocks the books. This means that in most cases you get same-day shipping if your order is received by 4:00 P.M., Monday through Friday, Central Time. But wait, it gets better. This simple aviation book paradise is completely searchable. Just search by title, author, subject, ISBN, or publisher, and enter your keyword(s). A handy description, price, and in some cases, the book's cover photo accompany the specific book's ordering page. Having trouble pinning down that perfect aviation book? Scan the main page for answers. Browse by categories, new reviews, top 10 titles, new titles, and more.

Oh, and let's not forget the book reviews. Robert Mark and Brian Jacobson, noted aviation industry experts, carefully review some of the books found in AvBook.com. Sometimes short, other times longer, these reviews are at a minimum always forthright. I think you'll find them a valuable selection tool.

Free or Fee: Free.

MyPilotStore.com

http://www.mypilotstore.com
e-mail: info@mypilotstore.com

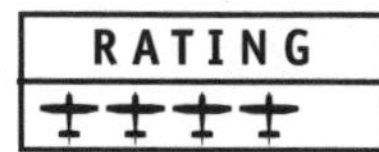

BRIEFING:

Well-organized online pilot supply store thoughtfully helps you buy aviation supplies and gifts.

Now, MyPilotStore.com knows how to sell aviation stuff. No, the company hasn't joined the ranks of the slimy used-car salesperson, but it does take a proactive role in selling aviation merchandise. And the best part is that the company's methods aren't annoying to the online aviation shopper, just helpful.

First a look at the site's presentation: It's streamlined for speed and perfectly functional. That's it. That's all you need to know. Navigation is near perfect, and the good stuff is usually no more than a click away.

Second, there's MyPilotStore.com's selection. It's endless. Categories fill the left-margin menu with such necessities as books and magazines, chart binders/accessories, collector signs, exam guides, flashlights, flight bags, gifts and toys, GPS, handheld transceivers, headsets/intercoms, software, training material, videos, and more.

Third, the true genius of MyPilotStore.com is in the extras. The Features section spells out your options: Chart Service, Gift Registry/Wish List (excellent handling of stuff you might want to actually purchase at a later date), Recommendation Center (helps you with gift ideas based on type of pilot), Affiliate Program (earn commissions by linking to MyPilotStore.com), Gift Certificates, and Aircraft Classifieds.

Free or Fee: Free.

Pilot Mall

http://www.pilotmall.com

e-mail: support@pilotmall.net

BRIEFING:

Vast database of aviation-only pilot supplies.

What would appear at first glance to be one giant online product database is in fact much more. Sure, the well-designed introduction page lists a few featured products coupled with a giant list of aviation categories and site search. Dig deeper to find a well-rounded online aviation shop at Pilot Mall.

The beauty behind this online beast lies in superb shopping-cart mechanics and your personal account management behind the scenes. Shopping is easy, and buying is even simpler. Just add your account information one time, and you're set. Next time you shop, just log in with a password. No need to reenter your contact information. The only thing you'll need to worry about is reigning in your overeager credit card.

Of course, the content is here too. Lots of products. Lots of categories. Lots of text links to get you to the good stuff. Choose from apparel, artwork by Sam Lyons, books and study guides, checklists, computer software, flashlights, flight bags, flight computers, forms, Gleim "red books," headsets, intercoms, Jeppesen, kneeboards, logbooks, plotters, tools and aids, transceivers, videos and tapes, watches, and more.

Free or Fee: Free.

Pilotbookstore.com

http://www.pilotbookstore.com

e-mail: pilotbookstore@charter.net

RATING
✈✈

BRIEFING:

Books, accessories, videos, and jobs? Pilotbook store.com serves up more than its name implies.

Owned and maintained by pilots, Pilotbookstore.com says what it is. It sells aviation books online. Oh, and it sells more than books. Confused? No need to be. Just realize before you click in that the company emphasizes books and dabbles in other things. Ultimately, the company's stated goal is "to promote aviation throughout the United States by offering discount aviation books, aviation supplies and accessories for low prices."

You'll find a lack of long product lists in text form or pull-down menus for that matter. Finding your needed novel or necessity is limited to a simple search box. At first, it seemed like I was flying under the hood. But type in a topic or product and you'll uncover the hidden booty. You'll find books, pilot test preps, training videos for pilots, pilot accessories, flight simulator software, and more.

As an added twist that really doesn't relate to the name at all, aviation job listings and a resumé posting service can be had at Pilotbookstore.com too. No, it doesn't really fit with the focus. But another opportunity to match pilot with jobs doesn't hurt.

Free or Fee: Free.

AvMart.com

http://www.avmart.com

e-mail: comments@avmart.com

RATING

BRIEFING:

Sheer online aviation shopping brilliance. A must for your favorites list.

AvMart.com easily breaks into the five-plane rating level. You know, where the brightest star performers are cruising along on the Web serving up aviation excellence. Every bit of AvMart.com creates a pleasant, perfectly satisfying experience.

If there was an online shopping model to follow—for any product or industry—it's AvMart.com. Page loading is immediate. Layout is perfect. Photos are tiny in size (until you get to a detail page of your chosen product). Help can be found everywhere (security information, quick search, and shipping/returns information). And the shopping cart with account information maintenance functions as it should.

The item catalog lists all the essentials: headsets, intercoms, GPSs and radios, pilot supplies (breaks down into tools, plotters, flight computers, logbooks, reference cards, kneeboards, flashlights, flight bags, aircraft, and more), training, books, charts and guides, software, videos, gifts and novelties, and apparel. Of course, as you would expect from online shopping excellence, featured products compete for your attention on the home page, too.

Frankly, there's not much more to say about AvMart.com. When a site is so simple to use and expertly crafted, the online experience speaks for itself.

Free or Fee: Free.

U.S. Air Salvage

http://www.usairsales.com
e-mail: contact@usairsales.com

RATING
✈✈✈

BRIEFING:

You could stumble on quite a find in this aviation salvage yard.

Congratulations to this Tennessee-based business that buys and sells single- and twin-engine aircraft. The company has managed to take its wares on the road, actually the cyber-highway, and provide the aircraft buyer or seller with this well-crafted Web symphony of ready aircraft, projects, rebuilds, parts, and more. The company's 20 acres with over 500 aircraft and a 20,000-square-foot warehouse of aircraft parts are behind the handsome presentation online. Shop for Cessna, Piper, Mooney, and some other oddball parts. Or dip into the nice selection of aircraft that's either ready to go or in the project stage.

The inventory, though changeable, demonstrates a nice variety, with such aircraft as a 1957 Beechcraft Travelair, a 1967 Cessna 210, a 1969 Bellanca Viking, a 1998 Mooney Ovation, a 1977 Cessna 310R, a 1954 Beechcraft E35 Bonanza, and many others. You get the idea. No jets so far, but no ultralights either. Just a nice selection of singles and twins. Each aircraft comes with a thorough description, photo (some with many), price, and contact form.

Free or Fee: Free.

Ooigui.com Airplanes for Sale

http://www.ooigui.com

e-mail: admin@ooigui.com

BRIEFING:

Oh, Ooigui.com, you've served up for-sale aircraft with simplicity and style.

Sort of like a flawless pattern and final approach lineup with stiff crosswinds, Ooigui.com Airplanes for Sale gives you that euphoric feeling of perfection with its aircraft-for-sale site. You'll be hard-pressed to find any site-specific acronyms or fancy Web jargon here. The site opts for the cut-and-dry approach. Best for those of us not interested in wasting a day on one Web site.

The machinery behind the site is built for speed and accuracy. Don't expect big, fancy graphics. Actually, the only pictures are of aircraft for sale, which is the way it ought to be. Menus are simple: Start (introductory page), Listing (sort all ads by date modified, aircraft, price, and state), Place an Ad (dealers, brokers, and private owners all have the opportunity to place free ads that stay up for 30 days), Wanted (add your own wanted notice on this bulletin board), and a nice list of aviation newsgroups in Talk.

I have no doubt that you'll appreciate the excellent presentation of aircraft for sale. The table is easy to understand, with headings of aircraft, date modified, state, price, airframe time, and engine time. The icons clue you in as to new ad with picture, ad with picture, new ad without picture, and ad without picture. Clever? Yes. Convenient? Absolutely.

Free or Fee: Free. Ads stay online for 30 days (as of review time).

WingsPilotShop.com

http://www.wingspilotshop.com

e-mail: online form

RATING

BRIEFING:

Cleared for online shopping, your final approach into WingsPilot Shop.com will be satisfying, rewarding, and stress-free.

Trying to earn your wings as an online aviation shopper? WingsPilotShop.com happens to be your "Top Gun" academy. Get out your checklist of needed items, and tap into that mental wish list too, because WingsPilotShop.com stocks it all. Just look at the product category list. The most extensive and specific I've seen in awhile, product searching includes aircraft models; aviator bears and accessories; license plates and frames; magnets and signs; pins; patches and key tags; posters, prints, and calendars; puzzles; seat cushions; toys; and more. Yes, the hardcore stuff's here too. You know, headsets, checklists, intercoms, transceivers, videos, test prep publications, and charts. Too overwhelmed? Just use the simple search. It's fast and painless.

Okay, you say, the content's there, but what about the site navigation? Is it easy to use? With frames o'plenty, which is not usually my favorite, product listings and product details appear dead center in the main frame. The categories are always to the left, and the main site menu is always at the top. So yes, site navigation is well equipped to handle your online shopping frenzy. Go ahead, break out the plastic.

Free or Fee: Free.

MyPlane.com

http://www.myplane.com

e-mail: service@myplane.com

RATING

BRIEFING:

My goodness, MyPlane. You've become a bookmarkable pit-stop during "MyNonFlying Time."

The first thing that strikes you at MyPlane.com is the "value added" tool box of extras. Yes, the company peddles planes here. You get some featured aircraft for sale on startup, for-sale searching, and lots of links to dealers. However, clear some hours on today's calendar for MyPlane.com, and have some snacks standing by. You may as well even pull the recliner up to the monitor. You might be here awhile.

Let's first assume that you're in the market for some new wings. Aircraft searching here is simple. Type in your keyword(s), or pick an aircraft category. Next, scan the details and view some pictures. Want the site to do the searching for you? No problem, just create an account (free) and submit your aircraft needs. MyHangar will notify you of new aircraft listings matching your criteria. You'll get your own personalized page for viewing aircraft. You can even choose to have updates to your page e-mailed to you as soon as they happen (up to once a day).

With MyPlane.com, extras are the norm. Not only will you find aircraft for sale, but you also will find worthy resources as well. Message boards, aircraft performance data, accident reports, ADs, dealers and brokers, aviation calendars, and owner clubs just begin to scratch the surface of site selections. See what I mean about the recliner?

Free or Fee: Most site information is free, but listing an aircraft for sale is fee-related.

Wencor West

http://www.wencor.com
e-mail: sales@wencor.com

RATING
✈✈✈

BRIEFING:
This commercial aviation parts supplier knows how to sell parts.

Since 1955, Wencor has been a commercial aircraft parts supplier heavyweight. As such, you'd assume that its online presence would be just as shiny in reputation. Sure enough, Wencor West sparkles with a solid Web application that meets its customer's needs perfectly.

Yes, its library of over 200,000 aircraft drawings and specifications, manufacturer's manuals, FAA advanced directives, advanced circulars, military technical standard orders, service bulletins, and OEM overhaul and repair manuals demands some credibility. However, doesn't it always come down to the quantity of the parts behind the pixels? Prepare to be impressed by the numbers: Wencor itself manufactures over 600 parts and is a stocking distributor of over 70,000 parts, with over 2000 customers, including over 250 airlines (all as of review time).

So, with all the great stuff behind the scenes, how's the online interface? In a word, *exceptional*. Design is appropriate, erring on the side of speed rather than beauty—which is just fine. Help and menus are everywhere. And functionality is top-notch. Attention commercial parts buyers: Wencor deserves a bookmark.

Free or Fee: Free, but you'll need to register.

AvionicsZone.com

http://www.avionicszone.com
e-mail: online form

BRIEFING:

Gizmo and gadget hunters take heed. Your avionics toy store is now online.

Hey, remember when Honeywell and Allied Signal merged in late 1999? No? Well no matter. Just feel confident knowing that AvionicsZone.com, a division of Honeywell, is tied to this $24 billion merged powerhouse. So really, you're looking at a megasite industry leader online.

This avionics wonderland (50,000 parts as of review time) specializes in new and retrofit avionics applications, primarily for commercial (air transport) aviation markets. Cool searchable databases include part numbers for product systems, installation kits, service parts, assemblies, components, and more. Along with the searchable databases, you'll find product information, online RFQs and ordering, a parts locator service, and pointers to factory-certified avionics repair stations.

Thinking about a little avionics update in your Cessna 150? Okay, bad example. However, for larger birds, look to AvionicsZone.com for traffic and collision-avoidance systems, ground proximity warning systems, communications and navigation systems, weather radar, flight voice/data recorders, and data-management systems.

Free or Fee: Free.

FlightGear.com

http://www.flightgear.com

e-mail: operations@flightgear.com

RATING
+++

BRIEFING:

Searching for that perfect headset, GPS, or avionics must-have? Push your cart no further than FlightGear.com.

AvGroup, the actual company behind the FlightGear.com name, stocks avionics, instruments, rotables, and accessories for corporate and regional turbine-powered aircraft. Even though its physical facilities happen to be in Atlanta, Phoenix, and Wichita, FlightGear.com is obviously well equipped to handle sales online too.

FlightGear.com will strike you as unassuming until you reach its substantial collection of pilot goodies in the Catalog area. You'll touch down smack in the middle of headsets, GPSs, software, books, training kits and videos, communications avionics, pilot supplies, emergency, and that forever-favorite aviation category, specials. Each category has many subcategories from which to choose. You'll be further encouraged when you find a wanted item. Click in for a nice description (sometimes very long) and excellent picture representation. Then check out using standard shopping-cart technology.

The product lines continue to grow, so stay alert. As of review time, the site supports most corporate and regional aircraft such as Cessna Citation, Beech King Air, Piper Cheyenne, and Jetstream 31/41, Hawker, Falcon, and Learjet by providing full lines of starter/generators, fuel controls and pumps, wheels, brakes, and pressurization controls and valves.

Free or Fee: Free.

The DC-3 Hangar

http://www.douglasdc3.com

e-mail: trev@douglasdc3.com

RATING

✈✈

BRIEFING:

Everything possible to know about the DC-3 is here and a bit more.

Every aircraft should have such a shrine as The DC-3 Hangar. It's no wonder that the lovable DC-3 gets its moment in the online sun for its part in aviation history. In the Hangar, you'd probably expect to see some pictures and historical write-ups. Yes? Well, hold on to your headset, this trip's going to be wild.

Obviously a volunteer effort, a major one at that, The DC-3 Hangar yields an unprecedented collection of everything DC-3 in one fell swoop. Although the design, grammar, site navigation, and spelling aren't on the A list for visiting, the incredibly vast assortment of information and pictures cries out for a bookmark. Take a breath first, and then scan my incomplete list of clickables: DC-3 Specs and History, FAA Data for DC-3, First DC-3 Flight, DC-3 Sounds, DC-3 Flight Simulation, DC-3 Checklist (cool!), DC-3's Oshkosh Photos (1997 and up), How to Fly the DC-3, DC-3 Forum, Stories from a DC-3 Pilot, Moving a DC-3 on the Road, and much more.

As much as this site's charm revolves around the volunteer, love-for-the-plane type of feel, big-time site characteristics slip into the content too. Get an instant report on the site's last update. Search the entire site with a keyword or two. And specifically search, sort, and display the collection photos based on your criteria.

Free or Fee: Free.

Aircraftrader.com

http://www.aircraftrader.com

e-mail: feedback@aircraftrader.com

RATING
+++

BRIEFING:

Aircraftrader.com lives up to its credo of "helping buyers and sellers connect."

Until you are an aircraft owner, you're just not equipped to throw up an aircraft-for-sale site. So it comes with some assurance to know that Aircraftrader.com grew from the collaboration of two aircraft owners "who know how hard it can be to buy or sell an aircraft."

They've obviously done their homework on what makes a site worthwhile. Aircraftrader.com provides online aircraft classifieds. Listings are free to browse, and the site is easy to navigate. Aircraft-for-sale listings are updated daily, and new listings are e-mailed out bimonthly to interested parties. Even aviation parts and services suppliers are listed.

Once you find your future wings, read the summary specs (year, price, equipment, contact information). Then click the camera for a usually large photo.

Free or Fee: Free to search, fee to place an ad.

AAR

http://www.aarcorp.com
e-mail: webmaster@aarcorp.com

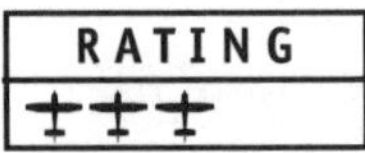

BRIEFING:

Well-designed parts searching, buying, and quoting.

For us small-time aviation types, AAR may be a bit intimidating. The well-designed home page opens many doors into parts buying, company information and alliances, AAR employment, investor relations, and more. Although AAR offers other industry categories (military/government logistics products and aviation commercial parts), nine out of ten readers probably will just want to pop into the general aviation new parts area.

The Web gurus behind the scenes obviously have worked tirelessly to create this comprehensive yet intuitive place for flying buffs to view parts, receive quotes, and purchase parts online. AAR's true online beauty lies in its dedication to perfect functionality. Registration is easy (although you'll have to wait for someone to call with a password). Browse your parts with simple searching. Set your own, custom-tailored user preferences. You'll never get lost with AAR's handy "breadcrumb trail" navigation. Track your orders online. And view your account information (account name, number, current credit limit, and remaining credit) on the left side of the screen.

Free or Fee: Free, but you must register first.

Trade-A-Plane

http://www.trade-a-plane.com

e-mail: webmaster@trade-a-plane.com

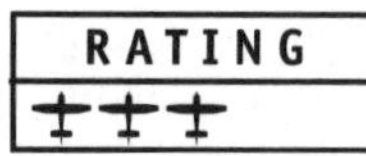

BRIEFING:

Industry print veteran dabbles online and perfects its wares on the Web.

Yep, you guessed it. This popular in-print magazine goes online too, but you'll have to be a subscriber to reap the real benefits. In the current age, where lots of quality online stuff is free to view, Trade-A-Plane bucks the trend a bit and requires a user name and password (get them instantly when you subscribe) to get to the good stuff. Don't get me wrong. Trade-A-Plane reigns high up there with the best aircraft-for-sale sites. So, in my humble opinion, the subscription fee's more than worth the benefit you'll receive from this credible resource.

As you dabble around, be sure to jump into Trade-A-Plane's broad menu offerings: Spec Sheet, Classifieds, Dealers/Brokers, Product Index, Advertiser Index, Forums, User Services, Event Calendar, Performance Database, Links, What's New, and more.

Penny-pinchers still looking for freebies will delight in knowing that aircraft-for-purchase searching is free (as of review time). Just search by aircraft type. You'll ultimately get a detailed preview of your prospective aircraft: photos, specs, price, description, and all the glorious tidbits that will make you drool.

Free or Fee:
Some information and searching are free, but subscription (fee-related) unlocks the good stuff.

Airbus Industrie

http://www.airbus.com

e-mail: online form

BRIEFING:

World-leading aircraft manufacturer secures its place as an online award winner too.

I suppose this book just can't stay current for too long. But at the time of review, the announcement of Airbus' future monster-sized carrier, the A-380, was just hitting the airwaves. So, as you would expect, Airbus had the A-380 plastered in every nook and cranny of its online promotional presence. With visual perfection, I might add.

Further enticing you into its Web of online information, Airbus Industrie online employs only the most useful, state-of-the-art site-development niceties that make a Web site reviewer smile. Exceptional, dynamic menus instantly pop up, with submenus when necessary. Subtle use of frames keeps your main site options omnipresent. Multimedia goodies are well organized and execute flawlessly (video clips, virtual reality QuickTime views, TV ads, vector drawings, and downloadable screen savers). And overall layout is kind of complex yet easily maneuverable.

Free or Fee: Free, unless you plan on snapping up a fleet of A-380s.

The Airport Shoppe

http://www.airportshoppe.com
e-mail: phoebe10@pacbell.net

RATING

BRIEFING:

Popular California-based supply shop serves up its wares online too.

Folding bikes. Now, there's something you don't see every day on aviation supply sites. Sure, you've got your headsets, books, videos, sectionals, and more, but bikes? Yep, The Airport Shoppe does have bikes. But that's really just the beginning. The cool folding bikes are just a hint at the variety of aviation goodies to be had.

Circle the pattern at The Airport Shoppe. Pages are refreshingly descriptive. Layout is quick and easy. Site navigation usually means you're just one or two clicks away from anything. Hundreds and hundreds of supplies are waiting in the wings. Most with pictures and detailed descriptions. With such an enormous bunch of products, The Airport Shoppe better have itself organized online. It does. Jump into four distinct categories to begin browsing: Oxygen, Supplies, Folding Bikes, and Books.

Once off the active and into a product-oriented section, begin clicking. A left-margin menu guides you like a finely tuned GPS to subcategories and more. Ready to buy? Good. Just click to buy, and a simple shopping cart takes you the rest of the way, securely of course.

Free or Fee: Free to view, but lots to buy.

Pilotportal.com

http://www.pilotportal.com

e-mail: service@pilotportal.com

BRIEFING:

More than a huge supply site, Pilotportal.com caters to aircraft club scheduling too.

Brimming with absolutely everything aviation, Pilotportal.com excels not only with quantity of stuff but also with quality. Most sites laying out this many products online would quickly become a confusing mess. Not with Pilotportal.com. A slick pull-down menu quickly categorizes and subcategorizes with ease. A right-margin product list and search box also team up to provide your Pilotportal.com flight following.

Along with pages and pages of supplies, descriptions, and pictures, Pilotportal.com stands out from the aviation supply crowd with its online bells and whistles. You can set up and monitor your own account with Pilotportal.com. View your "wish list" and order history, and change an address or password. Specifically, shopping is a marketing machine (in a good way) with an "Add to Wish List" button and a list of products purchased by fellow shoppers who also purchased the item of interest. Product reviews are also sprinkled throughout.

A handy little extra not found in most aviation supply sites is Pilotportal.com's FlightCalendar. With this cool tool, flying club members and owners in aircraft partnerships can easily schedule their aircraft usage through a secure, password-protected Internet link. Optional e-mail reminders of scheduled usage and online reports of aircraft usage help to keep users informed. It's fee-oriented but worth every penny.

Free or Fee: Free to use. Fee for FlightCalendar after 30-day free trial.

JETplane

http://www.jetplane.com

e-mail: info@jetplane.com

For-sale sites need only check here for a model of site aesthetics and function. JETplane site engineers earn accolades and a 300-knot flyby for this award-winning showing.

BRIEFING:

Expert design, organization, and professionalism take the active with JETplane. Don't forget your checkbook.

Prop planes, business jets, transport jets, helicopters, miscellaneous aircraft, and parts combine to form the mix of online classified ads. Yes, they're fee-oriented to list, but value seems obvious. If you're shopping, search by category. Scan the lists, use a keyword search, or just view the recent listings. Oh, and if you're the type who wants information to come to you (instead of the other way around), join the e-mail list to get notification of your specific product of interest.

Scanning the listing and finding your million-dollar turbine baby is painless. Narrow the search any way you desire (manufacturer name, year of manufacture, location, airframe hours, etc.), and move to a summarized list. Click an active listing to get the details: up to a 1000-word description, photo, and all the particulars. Then use the response form to contract the seller directly.

Hmmm, I wonder how much a G-IV is going for today?

Free or Fee: Free to search; fee to list ad.

South Valley Aviation

http://www.svconnect.com

e-mail: inquire@svconnect.com

BRIEFING:

Pilot supplies fast, affordable, and secure with online ordering.

C'mon, aren't we all looking for the best deal for pilot supplies? Headsets, books, flight computers, and other needed gadgetry can wreak havoc on the family budget. Enter South Valley Aviation.

Similar to the generic brand isle of your local supermarket, this "plain brown wrapper" site isn't flashy, but value is abundant nonetheless. Big-name pilot products litter these online isles. Start by selecting your category: books, charts, FAR/AIMs, headsets/intercoms, transceivers, GPSs, flight bags, flight computers, flight logs, flight-planning software, training, or videos. For search-savvy shoppers, it might be faster to just type your specific product with the search text box. Next, tap into pages and pages of "Internet coupons" worthy of a look to further discount your purchase. How many coupons? You won't be disappointed with the obligatory one or two. There are many, offering substantial savings on special products. There's no rhyme, reason, or expiration dates associated with the coupons, so check in often because they're changeable.

Finally, standard shopping-cart technology applies here. You know, select quantities, view the basket, check out, and so on. Transactions are said to be secure, so pull out the plastic and shop 'til your mouse hand drops.

Free or Fee: Free unless you buy something.

Helicopters Only

http://www.helicoptersonly.com

e-mail: mail@helicoptersonly.com

RATING

✈✈✈

BRIEFING

A nifty little one-stop chopper shop.

If the truth be known, Helicopters Only isn't only about helicopters. They've sprinkled a few fixed-wing goodies into this online supply stew too. However, the focus is helo stuff, and roto pilots should think "bookmark."

Relatively new but obviously growing into its goal of "your one-stop chopper shop," Helicopters Only begins your online shopping experience with a solid—not superlative—interface. A left-frame menu gives you the heads-up site display, while the main frame gets into detail. Initial site loading brings featured products to your attention, as well as some good training information. Read and click the product photos leading to headsets, book library (great selection of hard-to-find helicopter books), cockpit supplies, test prep guides and materials, gifts, and even some children's stuff.

Once you've spent your limit, check out with shopping-cart technology. It's simple, painless, and secure. By the way, it's encouraging to know that satisfaction is guaranteed at Helicopters Only. All products offered have been tested and approved by site administrators themselves.

Free or Fee: Free.

Cessna

http://www.cessna.com
e-mail: online form

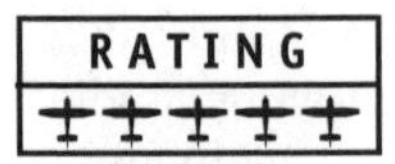

BRIEFING:

Model corporate site showcases its aircraft, company philosophy, and skill at public communication.

Stuffy? Dry? Humorless? Visually boring? No, not Cessna. This giant aircraft industry leader leans toward fun and fancy when it comes to its online showcase. Pushing my five-plane scale of excellence, Cessna creates the perfect corporate template of Web style. Employing every site mechanism known, the Cessna site gives you navigation perfection, rollover descriptions, pop-up windows, changing photography, and more. It's almost enough just to sort through all the gizmos. However, when you're done gawking at all the plane pictures and gadgetry, steer into a menu topic or two. Customer-building cheerleading comes in the form of "The Cessna Story." The good stuff, though, is found scanning the pictures and information on leading-edge aircraft. Read, see, and hear snippets on Citation aircraft, Caravan aircraft, and single-engine aircraft—all broken down further by model.

And when you're tired of scanning aircraft (no such thing for me!), there's much more taking the form of Cessna Jobs (employment opportunities) and In the News.

The Cessna site. It's a showy showpiece of Web mastery.

Free or Fee: Free.

The Mooney Mart

http://www.mooneymart.com
e-mail: info@mooneymart.com

RATING
+++

BRIEFING:
Interactive Mooney emporium delights, dazzles, and disseminates Mooney-only matters.

Free or Fee: Free.

If there is an inaudible way to be loud and boisterous, Mooney Mart's online presence has discovered it. Longwinded run-on commentary fills the pages, together with haphazard typos, clip art, sales pitches, and more. However, for the true Mooney enthusiast, you simply must make room for a bookmark.

Just take a deep breath and click in knowing that you'll be bombarded with the Web's equivalent to used-car sales propaganda. To provide appropriate content, the site prompts you to enter through one of three Web portals: existing Mooney owners, enthusiasts/prospective buyers, or general aviation industry vendors, dealers/brokers, etc.

Once in, everything Mooney may be found here in the form of sales, maintenance, modifications, forums, supplies, classifieds, links, and current events. Step into the introduction to get your feet wet with the five basic differences that Mooney Mart's site offers interactivity (Autopilot Al asks for information to personalize your online experience), Mooney-specific supplies, Mooney discussion groups and Q&As, free Mooney-oriented classifieds, and current events with Mooney Memories.

Mooney owners will be pleasantly surprised with automatic features such as Mooney Service Center SDRs (service difficulty reports), NOTAMS (notices to all "Mooniacs"), Mooney factory service bulletins, and FAA ADs.

List A Plane

http://www.listaplane.com

e-mail: online form

BRIEFING:

Classy for-sale site speaks volumes with organization and good design.

Departing the hordes of unseemly and unsightly aircraft listing sites, List A Plane rises out of the sales cesspool depths and rewards us with style and substance.

It's a perfect place to park your plane for sale. Menus, illustrations, and fancy site organization leave you feeling comfortable to sign up with these guys. Certainly, they claim prompt, efficient service. However, you'll have to evaluate this on your own. The easy listing format and examples guide you effortlessly into the arms of online sales. Even if you're on the buying end of the sale, you'll be equally impressed with helpful information request forms and a courtesy instruction section describing every search field. Search by manufacturer, model, N-number, aircraft age, price range, engine type, and more.

Based on your search criteria, a multitude of flying for-sale listings pop up with TTAF, SMOH, instrumentation, price, contact (e-mail and phone), and other detailed information. Some listings even serve up a photo or two.

Free or Fee: Free listings for a limited time (as of review time). Hurry, it may become fee-based someday.

Aircraft Dealer Network

http://www.aircraftdealers.net
e-mail: info@aircraftdealers.net

BRIEFING:
What's the big deal at Aircraft Dealers.net? Over 5000 listings, that's what.

Wheeling, dealing, and displaying, AircraftDealers.net may at first seem like that used-car salesperson who's immediately in your face when you pull into the lot. There are banners, multiple menus, and lots of announcements that seek your attention via various colors, styles, and sizes. However, hold firm and press on. AircraftDealers.net is loud (note the site's use of exclamation point punctuation) but thoroughly resourceful.

Your first clue to the resourcefulness of AircraftDealers.net is the duo of pull-down menus sitting atop the site's chaos. With only a click, you'll go directly to message boards, weather information, site mission, a quick tour, and lots of site advertising opportunities (including banners, business ads, and aircraft listings). Once you stumble across the aircraft search engines, you'll become entranced with the sheer volume of listings. How does over 5000 listings sound?

With such volume, the proper search engines aren't just nice, they are necessary. In this respect, AircraftDealers.net rises to the challenge. Search by aircraft, location, year of aircraft, and price range.

Free or Fee: Free to browse, fee to list your aircraft.

Pilot Toys.com

http://www.pilottoys.com

e-mail: controller@pilottoys.com

RATING

BRIEFING:

Well-named site stores the essentials for flight.

Ah, the name—how apropos it is. Except that the toys at Pilot Toys.com aren't of the stuffed animal or Tonka truck variety. Most, however, are just as fun to us pilot types. Fun has its price, though, and you'll find prices, products, and promotions o'plenty at Pilot Toys.com.

A modest yet workable interface brings you the whole online store, *sans* the bells and whistles. In fact, you can expect minimal photos, Javascripts, and ad banners that often can slow the searching progress. Take flight quickly with an immediate clearance for site navigation. A nondescript left-margin menu presents the choices, while a main frame displays the contents. Although the name is Pilot Toys.com, the company stocks mostly basics: test preparatory materials (books and software), gear essentials, CD-ROM references, textbooks and student kits, an aviator's library, and the new FAR/AIM series. Shoppers will receive good descriptions, adequate photos, and simple-to-use shopping-cart technology.

Free or Fee: Free.

Aircraft.CustomAds.com

http://aircraft.customads.com

e-mail: admin@customads.com

RATING

✈✈✈

BRIEFING:

Custom doesn't have to be complex. Find, sell, or trade aircraft easily online.

Emerging as a worthy resource is Aircraft.CustomAds.com—not necessarily for volume of listings but for overall site organization. Its automated systems make placing and updating ads simple. Expecting a complicated cockpit of controls? Grab the yoke and find only the essentials for easy flight. The topmost menu, which is eerily similar to browser menus, keeps you on a steady course with some site basics: Help, Search, Site Map (good!), FAQs, E-mail, and Home. Then roll into the free data havens of FAA Database Searches, Aircraft Sale Listings, Aircraft Trade Listings, Aircraft Wanted Listings, Classified Parts/Services Listings, and Dealers and Brokers Listings. Yes, there's a lot to search through here too. Become a member (free), place an aircraft sale listing, place an aircraft wanted listing, or place a dealers and brokers listing. Even review or update your ads.

Aircraft searching (if that's your game) is well organized with lots of narrowing fields: aircraft make and type, price range, ads listed within a user-specified number of days, and year range. Or find your flying dream machine based on N-number or airport ID search.

Free or Fee: Aircraft searching is free, and members receive one free, picture-less ad. Fee-related advertising is available.

AeroMall

http://www.aeromall.com
e-mail: none available

RATING
✈✈✈

BRIEFING:

AeroMall's stringent selection process makes these chosen aviation merchants the cream of the crop.

The old saying, "There's strength in numbers" certainly becomes evident as you stroll through the compilation at AeroMall. The difference? AeroMall maintains an affiliation of independent merchants who are handpicked and "approved as legitimate, time-tested operations." Yes, that's a big deal to the growing community of savvy aviation shoppers.

Although the pages are relatively attractive and organization is good, the real story is in the content. This one-stop-shop resource slips only the best and brightest aviation merchants in front of you. Curious about some of the requirements to be included on AeroMall? The merchant must have a professional, easy-to-use design, use industry-standard encryption (for credit cards), and display a telephone number staffed during normal business hours. Hey, now that's a beneficial selection!

So, although the rule here is quality over quantity, it's pleasantly surprising to see that the quantity is here too. Just browse the departments: Aviation Software, Books and Videos, Pilot Supplies, Aircraft for Sale, Aircraft Insurance, Hardware and Software, Luggage, Specialty, and Aviation Toys. Clickable merchants line the pages in an organized table offering at-a-glance information on each.

Free or Fee: Free, unless you're prompted to buy at a featured site.

Aviationgifts.com

http://www.aviationgifts.com
e-mail: rmdepar@uswest.net

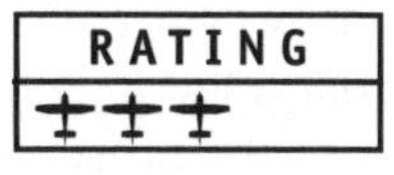

BRIEFING:

Fluff-free supply site makes shopping simple and secure.

For flight gear *sans* the fluff, step into Aviationgifts.com's stellar supply site. Relatively easy to use and devoid of time-wasting extras, this site skips directly to a huge menu of aviation gifts and necessities for sale online. Using the industry-standard shopping-cart system, Aviationgifts.com gets you through the store and the checkout line unscathed. Click and buy worry-free, with customer satisfaction guaranteed and secure server technology.

Although not a vision of perfect online beauty, Aviationgifts.com does an adequate job of presenting its wares. Organization and navigation are well designed with omnipresent menus and photographic icons. Item descriptions are succinct, and pricing is clearly marked on everything.

Categories from which to choose include apparel, novelties, outdoor products, mouse pads, books and videos, aviation software, headsets, handheld navigation tools, flight cases, aviation art, glasses, kneeboards, log books, plotters and flight computers, timers and watches, and more.

Free or Fee: Free.

AvShop.Net

http://www.avshop.net

e-mail: online form

BRIEFING:

No lines. No waiting. Step up and satisfy your pilot supply needs with online convenience.

Even though it's graphically oriented, AvShop.Net's online catalog loads fast. The layout people (who are obviously talented) must have meshed perfectly with the site-mechanics people because the final result exemplifies teamwork.

Once the initial welcome page instantly appears, you begin to sense that the interior will be just as fancy as the exterior. Click to enter the catalog, and you won't be disappointed, because lurking under the cowl of this aerodynamic machine beats the heart of a full-blown aviation supplies catalog. Mostly magnifying its software selections, AvShop.Net still provides a wider selection of necessities than you're led to believe. A nicely presented main menu moves you quickly into stuff like books and study guides, aviation videos, headsets, handheld GPS receivers, aviation gifts and toys, pilot training supplies, and a healthy dose of software specialties.

The site's beauty and brains also rise to the occasion during the "checkout." Once you've filled your basket with goodies, a well-organized system of ordering ensues. Review your basket contents, change or remove items, and proceed to either an electronic or offline checkout. It's easy and convenient.

Free or Fee: Free, unless you're tempted into ordering.

Spinners Pilot Shop

http://www.spinnerspilotshop.com

e-mail: comments@spinnerspilotshop.com

RATING

BRIEFING:

As online supply shops go, Spinners tops the list with revolutionary completeness.

Few pilot supply sites handle the complete online ordering system flawlessly. In fact, only a handful come to mind, with Spinners propelling its way onto my list. Layout and design work well together with little waiting time. Left-margin menu topics are always a click away. Online transactions are given secure routing. However, the best part revolves around Spinners' huge inventory. It's not the online mirage that most aviation supply sites throw together—just read some of their customer testimonials.

Without even touching the subcategory list, I'll give you a taste with headsets, Jeppesen, pilot supplies, books, logbooks, software, flight computer, flight bags, videos, training aides, kneeboards, flashlights, GPS systems, intercoms, handhelds, plane supplies, and more!

Shopping is relatively painless with the standard shopping-cart approach. Simply view product photos, read descriptions, review prices, and add items to your cart. Then check out online or offline. Simple instructions guide you through either method.

Get online, grab a mouse, find your favorite plastic card, and spin your way into a buying frenzy. It's easy to do.

Free or Fee: Free.

Plane-World

http://www.plane-world.com
e-mail: WebPilot@ComMaric.com

BRIEFING:

Kick the tires, rattle the flaps, and check for prop dings online with Plane-World.

All review criteria aside for a moment, the real test for aircraft-for-sale sites comes down to volume of listings. Selection is key, and to be frank, most aircraft-for-sale sites fall dramatically short in the listings area. Worse still, fly-by-night cyber-brokers tend to throw up visually forgettable Web tangles, expecting you to find your way through their meager offerings.

Rising above the scattered masses with Vx climb, Plane-World blasts onto the for-sale scene like a trusty Cessna 172 gone turbine. Appropriately simple menus and well-designed layout clear your way into a huge variety of classified listings. Following a left-margin menu that changes into submenus with each topic, page reckoning is of the 100-mile-visibility variety. Start with For-sale or Wanted. Then try the handy search engine or place your own free classifieds.

Although pictureless, the listings couldn't be more efficient and informative. The lack of pictures may put off some, but that's why the site's so fast. It's a tradeoff I'll take gladly.

Free or Fee: Free to browse and to add your own classifieds.

Aeroprice

http://www.aeroprice.com

e-mail: sales@aeroprice.com

RATING
✈✈✈

BRIEFING:

Holding your hand through buying or selling, this site's cool features make for a worthwhile stopover.

Buying and selling aircraft. In the lifetime of most serious aviators, the murky abyss of either endeavor may be riddled with uncertainty—especially the first purchase or sale. However, breaking through the low-lying fog of confusion, Aeroprice offers a progressive taxi toward understanding.

Although limited on any visual wizardry, Aeroprice provides some handy fee-related services, free tips and trends, and a gentle push toward additional resources. No, I'm not navigating under the hood. I realize that this for-sale site is subtly maneuvering toward cybersales of its QuickQuote online pricing and appraisal software. But it's certainly worthy of a flyby if you're shopping or selling.

After entering information into a thorough online questionnaire, you'll be introduced to an excellent aircraft pricing analysis. Look for great insight into pricing adjustments based on the average retail cost for your selected aircraft. Items analyzed include airframe, engine, avionics, additional equipment, interior, exterior, and damage history.

Free or Fee: Free.

Bombardier Aerospace Group

http://www.aerospace.bombardier.com

e-mail: none provided

BRIEFING:

Jet shoppers or just dreamers will find the ultimate in online perfection with Bombardier.

If you were searching for that flawless corporate aerospace site against which to compare others, Bombardier Aerospace Group's Web presence is the model. Organization and page designs are among the cutting-edge variety. Photo gallery images are professionally striking. And page navigation is effortless. Although a fully functional and informative gem, the site is simply a standard-setting masterpiece.

Okay, so maybe you're not in the market for a Canadair Regional Jet or a Challenger, but most enthusiasts will agree that the company's line of aircraft is worth an appreciative peek. Luxurious business jets like Global Express, Challenger, Canadair, and Learjet grace the company's online pages with specs and pictures. Similarly, regional aircraft information for the Canadair Regional Jet and the de Havilland Dash 8 series is at the ready.

Those more interested in employment rather than jet shopping will appreciate the site's extensive online personnel department (originating in Montreal). Detailed job offerings are categorized into Administration, Customer Support, Information Systems, Engineering, and Manufacturing.

Free or Fee: Free.

Microsoft Flight Simulator

http://www.microsoft.com/games/fs2000

e-mail: not available

BRIEFING:

Get the real briefing straight from the source—before you fire up those flight simulator engines.

The Flight Simulator site combines a couple of magical components inherent to all award-winning sites: valuable online information and easy communication to a huge audience (I bet you know a Microsoft Flight Simulator user).

Sure, there's software promo stuff everywhere. Ad banners and purchasing pages abound. But looking a little further into the mix, Flight Simulator users will strike virtual gold. The latest product information and downloads can be found in News. You'll find articles relating to flying, weather, scenery, and more. Performance specs, flight scenery enhancement products, and current Flight Simulator news are also handy for cyber-flyers. Online tech support is also standing by with customer support, product supports, FAQs, and steps to resource problems.

Whatever your simulator readiness, all Microsoft Flight Simulator users need a quick refresher here. And soon-to-be users? It's simply a perfect preflight walk-around.

Free or Fee: The information and tips are free. The software isn't.

Aircraft Shopper Online (ASO)

http://www.aso.com

e-mail: service@aso.com

RATING

BRIEFING:

It's an award-winning site for a reason. Unique searching capabilities smooth out online shopping turbulence.

Keeping the value of your time in mind (hey, you could be flying instead), ASO's pages are effortless and efficient. It's mostly text until you get to your selected destination. Once you plunge into countless aircraft-for-sale listings and narrow your search, you'll get descriptions and photos. Serious aircraft shoppers should skip directly to PowerSearch for excellent sorting and criteria setting. Set price and date ranges, limit the search to one aircraft make, or scan the entire listing. If you are (or will become) a frequent ASO shopper, these crafty Web developers have even included a clickable area for new ad additions and changes—showing only changes added in the number of days you specify.

Also handy are Dealers and Brokers, Financial Services, Want Ads, Letters to ASO, News Links, and i.ASO (personalized, custom system for finding your wanted aircraft).

Free or Fee: Free.

Air Source One

http://www.airsource1.com

e-mail: service@airsource1.com

RATING

BRIEFING:

Speed through this quick checkout line for student, corporate, military, and airline pilot supplies.

Grab an electronic shopping basket and stock up on your favorite aeronautical necessities with Air Source One. Simply put, you'll breeze through the quick checkout with no lines and everything you could possibly need.

This giant online pilot supply superstore expertly offers up isles of products from which to choose: headsets, GPSs, transceivers, charts, other electronics, FAA test preps, flight bags, aviation books, apparel, gifts, necessities, and software. Conveniently complete your order by credit card (the site ensures secure credit card processing), phone, fax, or mail. Online ordering can be next-day delivered and includes an e-mail confirmation.

If your basket gets heavy and you just want to find a particular item in a hurry, search by manufacturer or item description and simply type in your item. More minor site features include weather, company information, and other aviation sites.

Free or Fee: Free, unless you buy something!

Jeppesen

http://www.jeppesen.com
e-mail: online form

BRIEFING:

Artfully organized electronic catalog offers useful pilot supplies.

With well over 60 years of industry leadership (from a man who invented aviation charts), Jeppesen Sanderson once again captures the aviation world's attention with a visually captivating, organizationally brilliant Web companion. From the world's leading publisher of flight information (computer flight-planning services, aviation services, and training systems), Jeppesen's site provides a current online look into its offerings. Conveniently located among the company's history and profile information, you'll find the promotional core: the Jeppesen catalog.

While most "e-catalogs" weave the shopper through time-consuming pictures and tangled disarray, you'll glide effortlessly through Jepp's chart navigation and information services, aviation training courseware, NavData, software products, pilot supplies, and more.

Free or Fee: Free.

Wings Online

http://www.wingsonline.com

e-mail: wingsonline@cwtel.com

BRIEFING:

Aviation shopping is made easy with this site—crammed with specs and pictures of aircraft for sale, rent, or lease worldwide.

I'm always a fan of productive, visually appealing Web creations resulting from skill and a lot of elbow grease. The moment you grab the yoke here and finesse the controls, you'll also believe someone spent some late nights fine-tuning the many subtle nuances. This fantastic aircraft shopper resource gives tire kickers and eager buyers alike good information and many search choices. Sort by aircraft type, price, location, or the latest additions as of current date (marvelously efficient!). Search through related listings, such as For Sale by specific N-number and specific seller. Once you narrow your search and actually tap into the seller's wares, you'll be instantly informed with important specs: TT, STOH, SBOH, SMOH, registration, avionics, interior/exterior, price, contact, and more. Most listings also include up to multiple pictures showing off such areas as exterior, panel, interior, etc.

The best part for you net-savvy shoppers is that you'll actually find loads of quality listings—not a smattering of local rejects. When you're ready to really shop, try on this site—it fits perfectly.

Free or Fee: Free.

Optima Publications

http://www.pilotsguide.com

e-mail: webmaster@pilotsguide.com

BRIEFING:

This Optima site deftly guides you through its popular printed airport guides.

Okay, so maybe I've slightly narrowed the geographic scope with Optima Publications. But click through the Pilot's Guide Online and you'll see why this company has become an award winner. With a focus on California, southwestern, and northwestern airports, the Optima publications offer vital and current airport information.

You'll find convenient and thorough descriptions of product offerings, including Pilot's Guide to California Airports, Pilot's Guide to Southwestern Airports, Pilot's Guide to Northwestern Airports, Fun Places to Fly, Aeronautical Chart Subscription Service, and more. Click on any product category for instant prices and ordering information.

Scan through information about the Pilot's Guide (history and what it is) and topics relating to current subscribers (revisions service, customer service, etc.).

It's a simple, well-organized sales pitch for some excellent pilot products.

Free or Fee: Free.

AirShow–Aviation Trading Network

http://www.airshow.net

e-mail: crewchief@airshow.net

BRIEFING:

This expertly arranged for-sale site combines thoughtful features and a wondrous assortment of quality aircraft.

I urge you to discover the core of the AirShow's hidden talents. Just try looking up your favorite aircraft. While most aviation for-sale sites skimp by with only a few aircraft, the AirShow explodes with a huge variety of listings. It's a true resource for buyers and sellers. Clickable menu choices include Aircraft for Sale, Aircraft Dealers, Place an Ad, Change an Ad, and more.

The expertly developed searches within the huge database can be narrowed by price range, aircraft type, special characteristics, year range, and aircraft make/model. Serious plane hunters will rejoice at the brilliant What's New search—giving you only the latest additions since a user-specified date.

Not shopping, just selling? Well, you'll be equally impressed. All details for listing your unwanted flying machine are here—just click. Conveniently, you have a choice between an online ad form or custom service (mail, fax, or e-mail your photos and information).

Free or Fee: Free. If you're interested in showing an aircraft, reasonable fees apply.

Raytheon Aircraft

http://www.raytheon.com/rac

e-mail: online form

BRIEFING:

Professionally prepared corporate look at Raytheon Aircraft.

Originating from the Raytheon Company home pages, I invite you to skip directly to this nicely informative company page specifically devoted to Raytheon Aircraft. You'll quickly be enlightened about the company's broad product line. The nicely organized format gives you efficient descriptions of the Hawker 1000, Hawkers, Beechjets, King Airs, Bonanzas, Barons, and more.

Company information sources guide you to stuff like Events Calendar, Newsletters and Publications, Shareholder Information, Aircraft in the News, Aircraft Services, Travel Air, Raytheon Aerospace, and a What's New section. More clickable categories include Key Business Areas and Employment Opportunities. Still lost? A convenient site map makes finding a topic effortless.

Free or Fee: Free.

Boeing

http://www.boeing.com

e-mail: wwwmail.boeing@pss.boeing.com

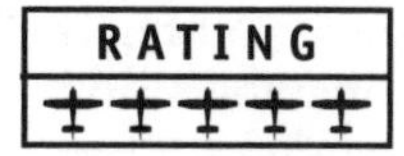

BRIEFING:

The world's leading commercial airplane manufacturer blasts off with more Boeing brilliance.

Among the aircraft manufacturers vying for some of your cyber-time, nothing Web-wide comes within a nautical mile of Boeing's online extravaganza. You could easily spend hours (even days) and not unearth every informative tidbit. Graphics, page navigation, pictures, facts, and surveys masterfully combine to create this interesting look into the world's leading manufacturer of commercial airplanes.

In addition to its leading manufacturing position, Boeing commands respect with its capabilities in (and informational Web pages relating to) space systems, rotorcraft, military airplanes, missile/tactical weapons, electronics/information systems, business jets, and associated products. Get an insider's peek into this jumbo company with a quick look at Boeing (at-a-glance information), News (financial, shareholder information), and feature Boeing-related stories.

Also wonderfully prepared are the thousands of information pages relating to your chosen tours (complete with photos) and vast employment opportunities. The employment "area" is a grand affair with subjects relating to college recruiting, internships, current opportunities, benefits, and submitting a resumé.

Free or Fee: Free.

The New Piper Company

http://www.newpiper.com

e-mail: none provided

RATING

✈✈

BRIEFING:

Piper product pitch pages captivate with nice pictures and performance specs.

To get something out of the New Piper page, you really don't have to own one or carry exclusive membership credentials from some elite Piper club. Even non-Piper junkies will enjoy browsing performance specs and pricing and comparing equipment lists on the entire currently manufactured fleet.

Aircraft choices for further examination include the Warrior III, Arrow, Seminole, Saratoga II, Archer III, Malibu Mirage, and the Seneca V. If the pages move you, a clickable map easily locates a dealer in your area. Or if you're more an occasional flyer than a buyer, get the scoop on company tours with a factory tour information page or past Piper press releases or browse pilot shop goodies.

During your stopover, be sure to visit the full-blown informational sections dedicated to the Malibu Meridian (cool single-engine turboprop). Lots of menu options and pages are devoted to the Malibu Meridian—Piper's newest single-engine baby.

Free or Fee: Free.

Lockheed Martin Corporation

http://www.lockheedmartin.com

e-mail: online form

RATING

+++

BRIEFING:

Lockheed Martin is light years ahead of its time in understanding how to appeal to aviation enthusiasts with the Web.

With ornate visual aesthetics riding shotgun, Lockheed Martin obviously sat the content people in the left seat for this highly informative site. Early on during this online flight you're inundated with a visual introduction into company contents. Lists of topics literally fill the pages under category headings of News & Announcements, Products & Services, Careers, Gallery, and Investor Relations.

You're looking for examples, aren't you? Remembering that these will change after review time, you'll be browsing articles like "New U.S. Air Force F-16s Will Have Color Displays and Other Advanced Systems," "Manned Space Systems to Produce Tanks for Reusable Launch Vehicle," and "Lockheed Martin Completes Initial Design Review for Its Joint Strike Fighter Program."

Once you've muddled through lengthy topics and summarized tidbits, do check into the image gallery—the resources are endless. There's a Lockheed Martin photo archive, video library, and many television commercials in QuickTime format.

Free or Fee: Free.

AircraftBuyer.com

http://www.aircraftbuyer.com

e-mail: dmperry@access.digix.net

BRIEFING:

A great aircraft "e-zine" that steers clear of "smoke and mirrors" information.

More than mere "e-zine" camouflage, A/C Flyer's AircraftBuyer.com dutifully avoids the cheap subscription-only tease and serves up meaty aviation delights. This flashy electronic version of its popular sibling in print may surprise you with helpful formats and regularly updated news.

Well-organized search engines guide you to endless listings of aircraft, products/services, and dealers/brokers. Current resale news and industry information are always at the ready, along with lots of flight training, financing, and insurance. And the informative ownership articles touch on many important topics. Aviation links, About AircraftBuyer.com, performance charts, and information on the Premium Club round out your thorough options.

Moving through the useful content is a snap. The subtle use of appropriate design elements balances nicely with useful navigation. Left-margin menu buttons help categorize your selections. And well-thought-out input forms guide you like a finely tuned GPS.

Free or Fee: Fee-based to list your aircraft, free to search for one.

PlaneQuest

http://www.planequest.com

e-mail: info@planequest.com

RATING

BRIEFING:

Aircraft buying, selling, and everything in between.

A cool, cockpit-style interface greets you on startup. From the get-go it will become obvious that PlaneQuest is organized, thoughtful, and right on target when it comes to buying and selling aircraft. Sure, you can search for and list aircraft for sale here. But the bookmarkable beauty is under the surface.

Dig into that left-margin menu for everything PlaneQuest has to offer: listing a plane, listing statistics, buying a plane, the selling process, the buying process, calendar, operating costs, discussions, and an excellent site demo. The selling and buying process information resources are really my site favorites. Fantastic in-depth primers on selling and buying a plane will leave you informed and in control—complete with the steps involved, pitfalls, things to ask, and so on.

Oh, and be sure to check out the unique Side by Side Comparison of Aircraft for Sale. This comparison (self-described as the only site offering such a comparison) gives the buyer the ability to evaluate aircraft for sale on 50 different categories in a spreadsheet format. The spreadsheet can be customized to show only the categories that interest the buyer. In addition, it can be used by a seller to determine the proper asking price for his or her aircraft based on the current market. Great idea! Thanks for the resource.

Free or Fee: Fee-based to list your aircraft, free to search for one.

The Flight Depot

http://www.flightdepot.com

e-mail: webmaster@flightdepot.com

BRIEFING:

Pull into the Depot for pilot supplies o'plenty.

Depth and breadth of products are usually what determines a supply site's success. It's no different with The Flight Depot. This site passes the test. The company adds in a few cute buttons, graphics, and menus for a solid overall performance. But let's focus on products.

For a feel for supplies at The Flight Depot, get into the left-margin menu, which changes into subcategories if applicable. Here you'll find a nice, manageable assortment of necessities. Accessories, books, charts, flight bags, GPSs, headsets, intercoms, kneeboards, logbooks, transceivers, and videos make up your main menu of options. Some are further broken down into brands or categories. Once you've found your category of choice, great product descriptions, photos if available, prices, and specifics fill your main frame.

As with most quality supply sites, standard online shopping-cart instructions apply here, too. Help, company information, and handy searching are also standing by for confused shoppers.

Free or Fee: Free.

Bookmarkable Listings

Advanced Procurement & Logistics System
http://www.apls.com
e-mail: sales@apls.com
Free access to current inventory, overhaul capabilities, and other aviation supplier reference information.

Rockwell
http://www.collins.rockwell.com
e-mail: online form
Corporate information regarding Rockwell's avionics, communications, and navigation products.

Avsupport Online
http://www.avsupport.com
e-mail: info@avsupport.com
Fee-oriented aviation parts searching.

U.S. Wings
http://www.uswings.com
e-mail: info@uswings.com
Manufacturer and distributor of aviation products.

WSDN Parts Locator
http://www.wsdn.com
e-mail: webmaster@wsdn.com
Aircraft spare parts locator and repair/supplier database.

Internet Parts Locator Systems
http://www.ipls.com
e-mail: webmaster@ipls.com
Database of aircraft spare parts and repair capability for the commercial aviation industry.

Beech Aviation, Inc.
http://www.beech-aviation.com
e-mail: info@beech-aviation.com
Beech aircraft-for-sale listings with photos, prices, and specs.

Nolly Productions, Inc.
http://www.nolly.com
e-mail: online form
Find videos, books, and software for training and career-related needs.

PartsBase
http://www.partsbase.com
e-mail: customerservice@partsbase.com
Click into this parts megamall for 24-hour shopping and no waiting.

Aerosearch
http://www.aerosearch.com
e-mail: info@aerosearch.com
Dip into Aerosearch's huge database when looking for tools and parts.

The Aviation Online Network
http://www.airparts.com
e-mail: webmaster@airparts.com
An easy-to-use, partly subscriber-based resource—great for finding parts.

Aviation Central
http://www.aviationcentral.co.uk
e-mail: webmaster@aviationcentral.co.uk
Buy and sell aircraft, aviation products, and services in the United Kingdom and Europe with free classified ads.

Stick and Rudder Pilot Shop
http://www.sandrpilotshop.com
e-mail: none available
Nice little supply site for discounted pilot goodies.

CloudTop Pilot Shop
http://www.cloudtop.co.uk/pilotshop.html
e-mail: info@cloudtop.co.uk
U.K.-based online collection of low-priced, high-quality aviation products.

Planes4saleonline.com
http://www.planes4saleonline.com
e-mail: shoot@cwcom.net
Site specifically aimed at the United Kingdom–European aircraft-for-sale market.

TheHangarFloor.com
http://www.thehangarfloor.com
e-mail: admin@thehangarfloor.com
"Real-time" middleman offers aircraft buyers and sellers a forum to meet online.

Avolo
http://www.avolo.com
e-mail: info@avolo.com
Use Avolo aviation marketplace for aircraft parts and service procurement.

Vector Quest
http://www.vectorquest.com
e-mail: sales@vectorquest.com
Pilot supplies, message board lounge, links, and more.

SellAircraft.com
http://www.sellaircraft.com
e-mail: online form
Great classifieds, links, pilot chat, and more.

Aerotrading
http://www.aerotrading.com
e-mail: info@aerotrading.com
Online marketplace for aircraft, products, and services related to aviation.

PilotSupplies Company
http://www.pilotsupplies.com
e-mail: info@pilotsupplies.com
Quality brand-name pilot supplies available online.

Blue Sky Flying Club
http://www.fly-bluesky.com
e-mail: replies@fly-bluesky.com
Online shopping for a wide range of aviation-related products.

Avid Aviator Pilot Supplies
http://www.avidaviator.com
e-mail: avidaviator@avidaviator.com
General aviation supplies with an emphasis on IFR flight simulators and computer-related aviation products.

Dassault Falcon Jet
http://www.falconjet.com
e-mail: see Web site for appropriate e-mail
Corporate information regarding the sales and support of the Falcon family of business jets.

Gulfstream.com
http://www.gulfstream.com
e-mail: info@gulfaero.com
Artfully produced online presence promoting Gulfstream aircraft, service, support, careers, and corporate information.

Embraer
http://www.embraer.com
e-mail: online form
Embraer, one of the four largest commercial aircraft manufacturers in the world, serves up its company and aircraft information with a stellar package of Web design perfection.

Internet's Best Aircraft Listing
http://www.bestaircraft.com
e-mail: contact@bestaircraft.com
Up-to-date listing of aircraft for sale on the Web.

GARMIN International
http://www.garmin.com
e-mail: webmaster@garmin.com
Giant corporate site devoted to all things GARMIN, complete with online store, product support, and lots of GPS information.

Aviation Entertainment

Airplanecards.com

http://www.airplanecards.com
e-mail: online form

BRIEFING:

Create online greeting cards for airplane, helicopter, and space enthusiasts.

As the name suggests, this site offers free airplane, helicopter, and space electronic greeting cards. Okay, it doesn't suggest the "free" part. But believe it or not, it is free. With over 700 images in its galleries (as of review time), Airplanecards.com is bound to fit the mood of your message. Although it's billed as "the Internet's first provider of free airplane, helicopter, and space greeting cards," you can never be sure about claims of "first, only, biggest, and best" on the Web. Rely instead on your own judgment. Try it.

The site design encourages participation in a friendly, easy-to-understand way. Good instruction and intuitive menus guide you through the free and fun process. Your first step involves picking a type of card (civilian, military, space, or hot air balloons). Next, you'll add recipient information. Third, choose a picture (there are many from which to choose). Fourth, add a greeting, message, and music. Finally, send.

Cheesy? Corny? Maybe. But I'll guarantee you this: Everyone receiving a card from me loved it nevertheless.

Free or Fee: Free.

13th Bomb Squadron Association

http://www.13thbombsquadron.org
e-mail: info@13thbombsquadron.org

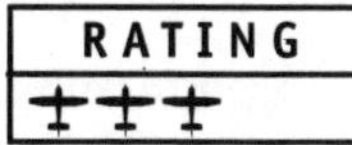

BRIEFING:

Words, pictures, and perspective make this one an award winner.

If the 13th Bomb Squadron Association doesn't meet your Web site design standards, you're missing the point. If, however, you fill with pride when reading the heroic, historic journals of our proud U.S. armed forces, then by all means pull up a monitor and start clicking. History and heroes abound here.

Such a complete tribute must be awe inspiring to members of the 13th Bomb Squadron. During the Korean War and including the period shortly thereafter, approximately 3000 men passed through the squadron. Even so, the chances are slim that you were among them. The 13th Bomb Squadron Association site sparkles with mass appeal, however. Why? The stories. The history. The people. Quite simply, it's a fascinating read.

Specifically, site options guide you into People, Planes, Our Scrapbook, History, Perspective, War Stories, "Oscar" Mascot, Poetry, Association Info, Reunions, Links, and more.

Free or Fee: Free.

AviationEvents.com

http://www.aviationevents.com

e-mail: info@aviationevents.com

RATING

✈✈✈

BRIEFING:

Searchable list of current and clickable aviation events.

Are you aware of every air show, fly-in, and seminar upcoming? Of course not. Neither is AviationEvents.com. But AviationEvents.com does make it pretty easy to find hundreds of aviation events around the world. Although magnetic in nature to us flying types, air shows and fly-ins aren't always well publicized, especially with enough lead time to make travel arrangements.

Enter the unassuming hero of AviationEvents.com. Compiled in one place online, hundreds of current and upcoming events are quickly at hand. The best part is that the site continues to update the listings frequently. Jump into this simple site and pick your event date range, state, and category. The database speeds you to a summarized listing of results. Get the event name, city and state, and the beginning/ending dates of the event. Next, choose your event by clicking the event name. Get the details and contact information instantly.

Current database categories from which to search include Aerobatic Competitions, Air Shows, Aircraft Accessories & Modifications, Conferences, Conventions, EAA Meetings, Events, Fly-Ins, Helicopters, Hot Air Balloons, Other, Races, Training and Seminars, Ultralights–Gliders–Soaring, and Young Eagles.

Free or Fee: Free to view, and listing an event is free, too.

Stormbirds

http://www.stormbirds.com

e-mail: webmaster@stormbirds.com

RATING

✈✈

BRIEFING:

Kind of like an A&E documentary on the life and times of the Messerschmitt Me 262 jet, only it's online.

Introduced during the final year of World War II, the Messerschmitt Me 262 aircraft blasted on the scene as the world's first operational jet fighter aircraft. So it would stand to reason that this noteworthy piece of aviation history would warrant its own hangar in cyberspace. Welcome to Stormbirds. The obvious love, admiration, and downright fascination with the Me 262 fill this tribute-type site with lots of remembrances, photos, and expression.

Specifically, delve into the Stormbirds Annex, CFII Me 262 Project, Stormbirds at War, Stormbirds Forums (substantial number of posts to view), Watson's Whizzers (USAAF jet recovery team information), Photo Recon Center (open-source repository of images), and more.

Although the actual Web site layout could stand improvement, the design is not really the focus of Stormbirds. What is? Content, content, and links to more content. And how better to keep up with pages and pages of great content than an updates page summarizing site changes by month? Site-wide searching and excellent before-you-click summaries also aid in site navigation. Final analysis: Stormbirds shines with organization and content. Give it a bookmark.

Free or Fee: Free.

Friends of Army Aviation

http://www.friendsofarmyaviation.com

e-mail: kmobrien@kmobrien.com

RATING

✈✈

BRIEFING:

Find your Army buddy with Friends of Army Aviation.

Mark and Kevin, your illustrious Friends of Army Aviation authors, welcome you on startup and explain their efforts and objectives. Their goal is to keep Army friends in touch with each other. Their site, Friends of Army Aviation, does just that, acting as a central locator. The current database of Army friends totals over 1000 as of review time.

Don't expect many bells and whistles. Simple menus serve as your GPS. Fancy graphics are nowhere to be found. And overall aesthetics fall into the ho-hum variety. But the heart of the site is the Army friends database. Sorted into alphabetical groupings, the list is easy and fast to use. Get fellow Army friends' names, e-mail addresses, notes, locations, and Friends of Army Aviation signup date.

More than just a database of Army friends, the site also provides a nice variety of extras. When a little bit of ground time arises, try Pilot Talk, Links, Photos, Aviation Humor, and Pilot Shoppe.

Free or Fee: Free.

Cool Wings

http://www.coolwings.cjb.net

e-mail: webmaster@coolwings.com

RATING
✈✈

BRIEFING:

An emerging selection of cool stuff in Cool Wings.

Cool Wings isn't the big aviation entertainment heavyweight you would hope for at first. Sure, it's a great start. The groundwork has been laid. The menu items are appetizing. Its content as of review time, however, lacks that satiating amount most aviators and enthusiasts crave. Bookmark it, and watch for the increasing evolution.

Current menu options take you into Forum, Classifieds (categories so far are airplanes, real estate, helicopters, accessories, and experimentals), Gallery, Videos, Top Gun (online game), Chat Rooms (general aviation, Air Force, and B-52 Stratofortress Association), and more.

So, until we see a dramatic increase in quantity of posts in the forums, classifieds, videos, and aviation photos, I'll invite you to check into Top Gun for a fun little online attack game while you wait.

Free or Fee: Free.

Flight Deck Simulations

http://www.meriweather.com/flightdeck.html

e-mail: jerry@meriweather.com

BRIEFING:

No need to jump into your local 777 simulator. Just click into Flight Deck Simulations and get pretty close to the real thing.

Prepare yourself for the true meaning of the word *detail*. No kidding. Have lots of refreshment nearby and make sure your household chores are finished before clicking into Flight Deck Simulations. It's fascinating for any pilot or aviation buff.

As you'd expect, you'll be introduced to the flight decks of several major commercial carriers. Examine every clickable panel, knob, switch, and button of the Boeing 777, 747, and 767. All is superbly defined and highly detailed. Flight Deck Simulations is truly an amazing resource. Can't imagine the time it took to develop this site. Or the time I spent playing around with it.

Virtual reality (360-degree views) of cockpits, acronyms and definitions, various site searches, and a What's New section complete your options. In addition to this detailed flight deck introduction, be prepared for an unnerving amount of pop-up windows and slightly confusing site navigation. It's bookmarkable nonetheless.

Free or Fee: Free.

FlyingToys.com

http://www.flyingtoys.com

e-mail: service@flyingtoys.com

BRIEFING:

Paper airplane fanatics unite! FlyingToys.com unleashes the power of paper.

Oh, I suppose FlyingToys.com's real focus is to sell flying toys. But that's not why it received an award-winning mention here. Sure, the paper airplanes, rubberband-powered gliders, kites, gifts, and books look fun and easy to buy online. If you've ever had an interest in paper airplanes, though, FlyingToys.com is your treasure chest of information and facts.

From the authors of the *World Record Paper Airplane Book*, the *Kids Paper Airplane Book*, and *Pocket Flyers* paper airplane book comes a bookmarkable resource for paper aviation fun. Use this well-designed site as your home base, and jump right into the "free stuff" of World Record (paper airplane world records, history, and Q&As), Aviation Humor, Airplane Club, Paper Flight Sim (fun little online paper airplane flying simulator), and Links. Of course, you'll also find a few free paper airplanes, complete with instructions, to assemble.

The site also includes an in-depth, *free* paper airplane class for teachers of all age levels. Just another excellent reason to put this one on your list of favorites.

Free or Fee: Free information, and lots of "toys" to buy, though.

The Aviation Humour Collection

http://www.aviationhumour.co.uk

e-mail: chris@aviationhumour.co.uk

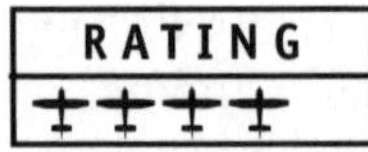

BRIEFING:

One of the best collections of aviation humor Web-wide.

After years of watching the Web and hoping, no yearning, for a solid aviation humor site, I've found my uproarious utopia. Yes, The Aviation Humour Collection fills the Web's aviation humor void that has seen many temporary sites offer a giggle or two and then disappear.

This new humor stronghold is no beauty by any means, but it's fast, navigable, and chock full of your soon-to-be favorite aviation jokes, stories, and dialogues. Reach into the Collective for a long list of clickables, such as Three Rough Landings, Is Nothing Sacred?!, Pilot's Prayer, Joke of the Week, Navy Pilot, The 33 Greatest Lies in Aviation, Cows Might Fly, Santa and the FAA, Hunting Season, and many more. Some of the jokes even offer little hints as to accompanying pictures or if they're among the best in the list (denoted by a series of yellow stars).

Free or Fee:
Gloriously free.

PlaneTalk7.com

http://www.planetalk7.com

e-mail: webmaster@pt7.com

BRIEFING:

"Cool enough to mix with us?" the site asks. Jump into the discussion group and show the participants that you are.

Under the "best kept secrets" category emerges the babbling communication forum PlaneTalk7.com. Although I hadn't heard of it until recently, PlaneTalk7.com seems to be quite popular. Maybe it's the aviation folks hopping around from discussion board to chat room and back. It seems that a surprising number of aviation chatters have landed here at PlaneTalk7.com with much to say.

An excellent system of menus and links begins your journey into discussion group euphoria. Scan the huge list of topics and posts. The numbers are staggering. Be assured, though, that browsing the list is relatively painless. Returning forum-goers need only view the left-margin "light bulb" to see the newest posts since their last visit.

Current forums (as of review time) include Air Forums, Old & New, Ground Talk Forums, World Talk Flight Attendant Forums, and Mixed Bag Talk Forums. Under these headings, find specific discussions that focus on health and safety, places to visit around the world, nonflying fraternity, cockpit rumors and news, and more.

Free or Fee:
You'll need to be a registered user to participate, which is free.

FlightProgress.com

http://www.flightprogress.com

e-mail: ggt14@aol.com

RATING

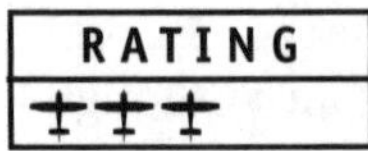

BRIEFING:

Flight tracking becomes skillfully efficient with Flight Progress.com.

FlightProgress.com allows anyone with Internet access to monitor the progress of specific airline flights in the United States and Canada without the hassle of calling the airline. After selecting the airline name and flight number, you'll see an aircraft symbol displayed on a large-scale map. Departure airport, destination, estimated time of arrival (ETA), altitude, and airspeed also are shown. And are you ready for this? Get flight arrival notification delivered straight to your mobile phone, pager, PDA, or desktop computer—*free*. Cool, huh? Don't really have a flight in mind, but you still want to check it out? No problem. Just click the Random Flight Tracker for an example of how the thing works.

The information is linked from FAA radar displays and is much more accurate than the information provided by the airlines. While you're aboard FlightProgress.com, be sure to check out the links to weather information, airports, and live audio from actual Air Traffic Control frequencies.

Free or Fee: Free.

Skytamer Images

http://www.skytamer.com

e-mail: john@skytamer.com

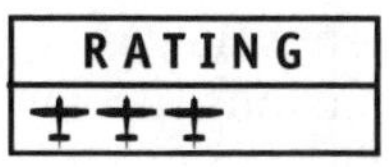

BRIEFING:

Aviation images that are anything but tame.

Casting aside the less significant introductory site options of Sea and Land, you'll doubtless want to jump into Air, as I did. Inside, you'll find a world of excellent aircraft photography worthy of your attention, downtime, and bookmark.

Rivaling some of the larger aviation reference sites, Skytamer does tame the burden of uncovering information and photos on your own. The site features an extensive aircraft photo collection (a long, long list of aircraft), a slide show, and aviation databases. Skytamer databases include USAF serial numbers, civilian aircraft registration numbers (N-numbers), U.S. military aircraft tail codes, aircraft specifications, an air show calendar, a glossary of aviation terms and acronyms, conversion factors, the international phonetic alphabet, and more. Its Museum Guide provides names, addresses, phone numbers, and active Internet links to aviation museums throughout the world (handy if you're a world traveler).

Oh, and Mystery Plane checks your knowledge of aviation history. Take a look at a tightly cropped aircraft photo and guess what it is.

Free or Fee: Free.

Avitop.com

http://www.avitop.com
e-mail: online form

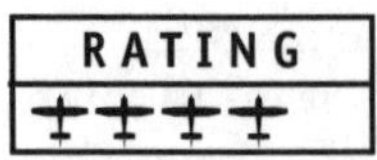

BRIEFING:
With obvious concern for satiating your need for quality aviation entertainment online, Avitop.com tops most similar sites.

Welcome to the Web's aviation entertainment smorgasbord: Avitop.com. Unlimited activities await your clicking fingertips. Getting down to some serious diversion here, Avitop.com avoids the boring facts, the FARs, and the forecasts. Replacements take the form of more trivial but worthwhile aviation endeavors, such as jokes, photos, trivia, N-number searching, and aircraft for sale. So, yes, it's a mixed bag filled with fun.

Certainly taking the high road to reach this aesthetically picturesque page layout, Avitop.com pleases the eye and the time-sensitive browser with equal balance. Its magazine-style presentation performs perfectly as Avitop.com shows off its wares. Menus are well defined. Navigation is clear. Graphics are optimal size. And content ties the whole thing together.

The Forum is particularly interesting, with topics ranging from Aviation 100 Questions to Questions about the F16. A substantial amount of posts fill the Forum, too. Not just a tiny smattering of queries.

Free or Fee: Free.

Air Warriors

http://www.airwarriors.com

e-mail: JohnWickham@airwarriors.com

RATING

✈✈

BRIEFING:

Peek into the world of the student naval aviator.

Inspired by ENS Dave Werner's online Journal of a Student Naval Aviator and a book entitled *Air Warriors: The Making of a Navy Pilot,* this Web site's authors capture the essence of naval aviator training online with compelling honesty and real-life experience.

Although it grew from just some good aviation jokes and a personal online journal, Air Warriors today has evolved into a full-blown naval aviator training handbook of sorts (unofficial, of course). Complete with an Updates Section, Forum, Quiz (database of flight training questions and answers), Gallery (lots of quality action pictures), Aviation Humor and True Stories, Biography, Links, FAQs, and more.

Probably one of the most beneficial areas to visit is the database of flight training questions and answers. Here you'll scan a long list of Q&As dedicated to operational limits, general aircraft, electrical system, NAVAIDs, and systems.

Free or Fee: Free.

Fighter Planes

http://www.fighter-planes.com

e-mail: Werner.Bergmans@iae.nl

RATING
✈✈

BRIEFING:

Toss the dusty history books and relish the "pixelized" perfection of these wondrous fighter planes online.

How in the world could I be involved in a book about the best aviation Web sites and not include a site solely dedicated to fighter planes? The answer is, I can't. So I've managed to locate Fighter Planes—a new favorite of mine spotlighting popular military flying machines. I thought I'd pass it on to you (and your bookmark list).

On this site you can find over 85 different fighter planes (as of review time), complete with technical information and pictures. Just click on the date range of manufacture (the fighters are indexed by manufacture date), and jump into the summaries. More detail and a full-sized image are waiting in the wings should your thirst for knowledge overtake you.

Curious about content? Come face to face (at least through your monitor) with aircraft like the MiG21 Fishbed, F104 Starfighter, F8 Crusader, B52G Stratofortress, EF111 Raven, F18 Hornet, Panavia Tornado, BF-109 Messerschmitt, Hawker Hurricane, and the A6M Zero.

Free or Fee: Free.

The Flying Clippers

http://www.flyingclippers.com

e-mail: comments@flyingclippers.com

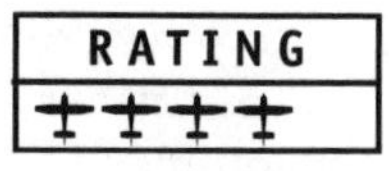

Briefing:

The romantic time of Pan Am's flying clippers takes flight online.

Pan Am's fabulous flying ships. Tough to find a more fascinating, romantic time in commercial aviation history than aviation's Golden Age. Luxurious, state of the art, and grand, these magnificent flying ships captured the interest of travelers for good reason. The great Pan Am Clippers—the Sikorsky S42s, the M-130s, and the B-314s—were probably the most romantic planes ever built. Reach into the history and live the day of the Clipper online with The Flying Clippers site. It's an in-depth, well-written journey with lots of photos, timelines, history, and more.

When your quest for Flying Clipper nostalgia takes a more tangible turn, step into the Post Flight Shop at Flying Clippers for books, videos, posters and art prints, aircraft models, toys, and more. It's memorabilia of aviation's Golden Age on sale now.

Free or Fee: Free.

Chuckyeager.com

http://www.chuckyeager.com

e-mail: klarsen@jps.net

BRIEFING:

One fan's historical perspective and collected history of General Chuck Yeager.

The amazing life and times of an American hero reside here at Chuckyeager.com through the efforts of this dedicated site's author. Expertly crafted, this online tribute details the lifetime achievements of Chuck Yeager. The site features an array of history, photos, videos, and timelines dedicated to this remarkable man.

In the site you'll learn that General Yeager has flown 201 types of military aircraft and has more than 14,000 flying hours, with more than 13,000 of these in fighter aircraft. He has most recently flown the SR-71, F-15, F-16, F-18, and the F-20 Tigershark. Fantastic video clips enrich your online journey. Your four options include a rare 30-second clip from 1947, the launch of the X-1 from a B-29, several air-to-air views of X-1 Number 2 and its modified B-50 mothership, and a clip showing the X-1 landing on Rogers Dry Lakebed followed by the safety chase aircraft.

Chuck Yeager made his last flight as a military consultant on October 14, 1997, the fiftieth anniversary of his history-making flight in the X-1. He observed the occasion by once again breaking the sound barrier, this time in an F-15 fighter.

It should be noted and understood that this unofficial fan site is in no way endorsed or sponsored by General Chuck Yeager.

Free or Fee: Free.

Skychick

http://www.skychick.com
e-mail: skychick@skychick.com

RATING

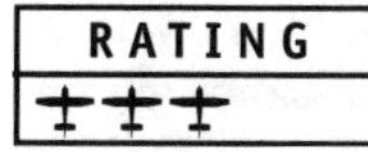

BRIEFING:

Single-handedly, Skychick tops the flight attendant bookmark list.

Well, if there was ever a quintessential one-stop shop for everything flight attendants could need, Skychick is it. Not just the practical stuff either. The humor, wit, and candor come in handy, too. In fact, the humor here provides that award-winning boost to keep surfers and servers entertained.

Skychick, self-described as "the definitive expert in the industry," caters to fellow flying attendants with this well-developed online show. The clickables as your virtual serving cart rolls by include Flight Attendant Humor, Articles (flying and families, how to get hired, surviving training, airport codes, uniforms, packing secrets, passenger rant, and more), and a few interactive forums.

Without question, the humor section should be your first stop. The fun prevails with Galley Giggles, Flight Attendant Prayer, Flight Attendant Songs, Slang Dictionary, Flight Attendant Horror-scopes, Are You a SkyBag?, Overheard/Flights, and Who Am I? Even if you're not among the ranks of the sky attendants, you'll appreciate the humor anyway.

Free or Fee: Free.

SR-71 Online

http://www.sr-71.org

e-mail: webmaster@sr-71.org

RATING
✈✈✈

BRIEFING:

Overflowing archives of SR-71 Blackbird photos, specs, and history.

Unofficially nicknamed the "Blackbird," the SR-71 was developed as a long-range strategic reconnaissance aircraft capable of flying at speeds over Mach 3.2 and at 85,000 feet. Today, it speeds through cyberspace too at SR-71 Online. Although it flames out when talk turns to site navigation (get ready to use your "Back" button or follow "bread crumb" trails), SR-71 Online gracefully flies on anyway to the delight of everyone.

Its more expanded title, SR-71 Online: An Online Aircraft Museum, hints at more that just a personal photo collection. The site harbors an information-rich archive of mostly Blackbird-oriented stuff. But other military aircraft information continues to build too. Primarily, SR-71 Online dedicates itself to information, photos, and diagrams regarding the A-12, YF-12, and SR-71. Rely on the home page as your jumping off point into site contents, such as Blackbird Archive, Military Aircraft, Groom Lake, Image Gallery, News, Links, SR-71 Online Store (I guess every site's got one nowadays), and Contact Information.

Free or Fee: Free.

Fly-ins.com

http://www.flyins.com
e-mail: admin@flyins.com

BRIEFING:

"Dedicated to the pure fun of it all," Fly-ins.com is your *TV Guide* listing for national and international fly-ins.

I think we can all agree with Fly-ins.com's contention that fly-ins and other aviation gatherings make up the heart and soul of aviation. Who wouldn't like spit-shined relics buzzing around and fancy new prototypes stirring our fantasies? Right. Everyone likes them. So you'll understand Fly-ins.com's popularity. It seems that many before me have discovered it. Just search the event database. There's a long list of events spanning the globe and the calendar.

Searching for your favorite type of event or date is easy, though. Just click into the Find an Event query form. You'll narrow your search with criteria such as categories (fly-ins, military, museum events, air shows, races, gliders, seminars, and more), date ranges (from when to when), and/or state.

Once you've uncovered an event that tickles your fancy, click for more details. Get the skinny on who's sponsoring the event, a description, dates and times, location, and contact information.

Free or Fee: Free, even to post an event.

AIRTOONS

http://www.airtoons.com
e-mail: online form

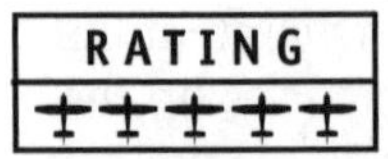

BRIEFING:

Imaginative but adult-oriented journey into the "hidden" meaning behind those pesky airline safety cards.

Be forewarned from the get-go: AIRTOONS isn't your mild-mannered, straight-laced online aviation comic strip. Those who are easily offended may seek humor elsewhere (although there is an 18-and-under section you could run to). With four languages (English, French, Swedish, and Hebrew) from which to choose, AIRTOONS takes its humor to many countries, not just English-speakers.

Ready for the 'toon review? First, a little background. From this illustrious site author's mischievous mind comes AIRTOONS, imaginative adaptations from airline safety cards (complete with captions). If you look closely, you'll notice that there are no captions for the imagery in your favorite commercial carrier's safety cards. AIRTOONS has taken it on itself to "discover the real meanings and post them here for all to see."

Funny 'toons, sometimes outright hysterical, are at the heart of this well-crafted site. Beyond the imagination, wit, and clever humor, the site is itself a masterpiece of perfect navigation and design. So sit back and enjoy the AIRTOONS ride. Just make sure you know where the exits are.

Free or Fee: Free.

The Plane Spotting Network

http://www.planespotting.net

e-mail: online form

RATING
✈✈✈

BRIEFING:

Do you look to the sky whenever the faint sound of jet engines can be heard? Click into The Plane Spotting Network, and thank me later.

Brought to you by the same talented author as at AirDisaster.com (also an award winner in this book), The Plane Spotting Network boldly attempts to sift through and present only the best airport information and aircraft pictures around.

Collected from people who live and breathe aviation, pictures and information pack the pages of The Plane Spotting Network. But before you get the impression that site navigation is unruly, trust me that competent, easy-to-use databases sort through everything beautifully. And there's a lot to sort. Beyond the long list of categories and searchable airliner photos, you have many options in the left-margin menu. The Spotter's Guide also brings you face to face with airport information, aircraft identification, airline fleet lists, N-number search, and spectacular airports. There's even a Spotters' Forum and Live Chat.

Free or Fee: Free.

Flyforums.com

http://www.flyforums.com

e-mail: webmaster@flyforums.com

BRIEFING:

The sense of "aviation community" is alive and well, due in no small part to aviation forums like Flyforums.com.

Flyforums.com is the spinoff from a now-defunct home project Web site (Hughesworld) focusing on the lively world of aviation forums. I'm sure you'll agree that on first impression the stellar site design certainly grabs you and creates instant respect. But it's what lurks behind the scenes that'll keep you coming back: the all-powerful and dynamic message forum.

Currently, your collection of forums includes ATC/NATCA, American Airlines & American Eagle, Continental, Continental Express, Delta/Regionals, Fractional, General Aviation, US Airways, and more. Once you enter any of the forums, organization and simplicity rule. Usage is perfectly intuitive, and posting is simple. Yes, but what about quantity of posts, you ask? Get ready for lots to read. Hours and hours of aviation interaction await.

Free or Fee: Free.

The Able Dogs

http://www.abledogs.com
e-mail: jerry@abledogs.com

BRIEFING:

An able tribute to The Able Dogs.

Flying and stories just seem to go together. Pilots like to share experiences. It's no different at The Able Dogs. Slip into this tastefully crafted site of tributes yourself. Specifically, the site is dedicated (you probably guessed it already) to the venerable Able Dog (Douglas AD Skyraider) and the men who designed her, flew her, maintained her, and otherwise had a hand in her naval aviation career. So you can imagine all the remembrances this site attracts.

The left-margin menu steers you into all the site embraces: The Skyraider (history, photos, schematics, and specs), What's New (excellent chronicle of site changes, complete with links), Sea Stories, The Dog Pattern (online forum), Image Galleries, USN Squadrons, Naval Air Patches, Reunions, In Memory Of…, and more.

Be sure to curl up by the monitor when you get some time alone and read some entries in the Sea Stories section. Captivating stories and poems are many. Selected favorites include "Old Yellow Stain," "Last Flight North," "'Horse' and The Dog Pattern," and "A Dark and Stormy Night."

Free or Fee: Free.

Plane Writing

http://www.xs4all.nl/~blago/planewriting/index.html

e-mail: blago@xs4all.nl

BRIEFING:

Plane Writing—your first, last, and only online stop to reminisce about the bygone era of prefifties flying.

Wow! Top honors in the historical/entertainment categories go hands down to Plane Writing. Just moving through its stellar organization, I clicked along in wondrous delight. And the content? Yes, the featured selections of vintage writing are better than any I've seen. Specifically, you'll become entranced with fascinating quotes from vintage writing about flying and early pilots' biographies.

It may almost be a disservice to describe the perfect style, artistry, and countless hours of organization obviously spent compiling Plane Writing. Suffice it to say that frames, multiple menus, and text-link summaries play a role. But you simply must experience it yourself.

Dive nose first into an array of quotes and story snippets from prefifties flying and living. Read about biographies and life from these heroes of an older era—one of exciting, new discoveries. This site takes you by the hand and subtly introduces you to history. Main subsections take you into separate long ago realms (each with its own menu of remembrances). Browse through selections entitled, "On Leaving," "Cities and Green Hills," "Scraps and Pieces," "WWI–WWII," "Other Metaphors of Flying," "Sounds and Smells," and "Farewell to a Plane."

A stopover at Plane Writing? They don't allow me enough room in this book to describe the reasons why. Just do it.

Free or Fee: Free.

Patty Wagstaff Airshows

http://www.pattywagstaff.com

e-mail: online form

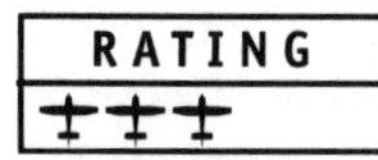

BRIEFING:

A dizzying assortment of Patty's pictures, planes, and people.

If you're any part the true aviation enthusiast you claim to be, you must've strained your neck on more than one air show by now. How can you not get caught up in the flybys, aerial stunts, and big radial engine static displays? Well, something tells me Patty Wagstaff enjoys air shows, too. Her Web presence, Patty Wagstaff Air shows, is an obvious indicator that her organizational skills don't end at air shows. She's learned a bit about Web design, too.

Sure, the site sacrifices lengthy loading times in favor of glitz, but the compromise seems appropriate. We are talking air shows after all. The photographic picture gallery grabbed my attention first. Lots of pictures. Lots of galleries. As you'd expect, her BFGoodrich Aeorspace Extra 300S fills the frames. But you may not expect the QuickTime movies of Patty in action—twisting, turning, and smoking. Have patience with loading times, though, it's worth the fun.

Continue on through Patty's page and you'll read about her background, meet the staff, and have her next airshow schedule at hand.

Free or Fee: Free.

Navy Flight Test

http://flighttest.navair.navy.mil

e-mail: online form

RATING

✈✈✈

BRIEFING:

Navy Flight Test joins the military photo fray with a visually stunning selection.

Perfect visual mastery begins your Navy Flight Test journey into resources and photos. Just keep in mind that the resources fall into the technical category of flight testing. You know, everyday stuff like mechanical systems, propulsion, mission and sensor systems, ship suitability, weapons/stores, and more.

Yes, I'm more into the photos, too, but the resources side deserves a bit more acknowledgment. Employment opportunities, flight test and safety lessons noted, weather, phone directories, and others also grace the pages of Navy Flight Test. Almost everything's way over my private pilot head, bit it's still interesting.

When you've grown weary of too much of the vague and technical, skip over to an all-star cast of quality photos in the flight test gallery. All searchable photos are courtesy of the Patuxent River Naval Air Station Photo Lab. While most photo-related Web sites skew toward the amateur variety, this one is top-notch. Use the simple search for proof. Scan through pictures of an F/A-18 Hornet, C-12, C-130, AV-8B Harrier II, F-14 Tomcat, HARM missile, aircraft carrier, and more. Even view by bomb drops, in-flight, missile shots, or ship operations.

Free or Fee: Free.

B-17 Flying Fortress–A Virtual Tour

http://www.b-17.com

e-mail: online form

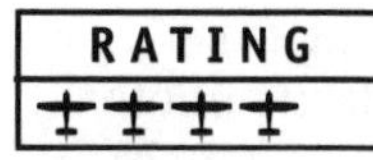

BRIEFING:

This site's flawless execution rewards the QuickTime viewer with up-close B-17 nostalgia.

Virtually spectacular in concept and execution, B-17 Flying Fortress offers this online tribute to the nostalgic B-17 superhero. Obviously, Web designers had fun in creating a visually succinct and inviting tour of this particular flying fortress physically located at the Lone Star Flight Museum in Galveston, Texas.

As you click in, you quickly discover that multimedia gadgetry was built into the source code. Don't worry, though. You'll just need to be familiar with QuickTime VR panoramas and QuickTime VR Object Movies. Need a refresher? It's only a click away.

Once you've found your bearings, five panoramas take you inside World War II's most famous bomber. Sit in the cockpit. Peer out the top turret. Be a bombardier. Use the radio. Or walk around the B-17, virtually, seeing every outside detail.

When you've become dizzy with too much multimedia video, check into the story behind the plane, the stats, and photography. It's virtually priceless.

Free or Fee: Free.

Airshow Action Photo Gallery

http://www.steehouwer.com

e-mail: peter@steehouwer.com

RATING

+++

BRIEFING:

Quality air-show photos abound courtesy of Peter.

Oh Peter. So many air-show photos, so little time. Thanks to an enormous collection of fantastic photos by Peter Steehouwer, our ground time is that much more palatable.

First, a warning about the objectionables: typos, layout, and organization fall into the seriously questionable category. However, once you've found the photos, you become lost in smoke, sky, and high angles of attack. Peter has captured the thrill of speed and precision with an endless selection of high-quality stills.

Just about every major worldwide air show somehow finds itself represented here with a collage of photographic reminders. From California to Arizona to Texas and from Switzerland to the Netherlands to the United Kingdom, Peter's Airshow Action Photo Gallery is a sight to behold. And with Peter, quality is obviously key. Proudly, he reports that all photos are aircraft in action, not ground shots. He shoots from ground to air, climbing atop anything nearby (car, mountain, tower, step ladder, etc.) with some pretty powerful camera equipment. Just look at his quality. Peter's skill is obvious.

For frequent visitors I offer this hint: Just skip directly to the scattered "New" tags and leave your e-mail address to be notified when the site is updated.

Free or Fee: Free.

Airshowpics

http://www.airshowpics.com

e-mail: online form

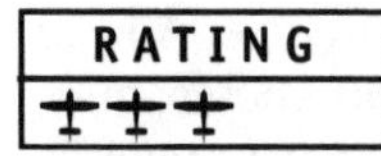

BRIEFING:

Massive collection of aircraft pictures leaves you utterly dazzled.

Loaded with more pictures than even hard-core collectors would own, Airshowpics completely dominates the very visual world of online air-show pictures. As expected with such a massive collection, quality of photos varies widely, but most fall into the crisp and captivating category. With Airshowpics, you need only sit back and click through the long menu of aircraft for a multihour visual show.

Featuring both flying and static aircraft on display, Airshowpics pumps adrenaline through your mouse and brings you up close and personal with the world's favorite flying machines. Begin by browsing the left-margin list. Display teams, fighters, bombers, airliners, freighters, helicopters, vintage, and special duty carry a collection of photos within. Actual selections within these categories include Blue Angels, Snowbirds, Turkish Stars, F-16, Tornado, Dassault Breguet, B-52, Concorde, C-130, WWII bombers, and reconnaissance aircraft, to name a handful.

Even when you've exhausted everything listed, go to the aircraft image archives for over 450 pictures or the great collection of movie clips. It's almost online overload.

Free or Fee: Free.

Paris Air Show

http://www.paris-air-show.com

e-mail: none available

RATING
✈✈✈

BRIEFING:

Get caught up in the "show of shows" with this online prep to the Paris Air Show.

Ah, Paris. The lights. The history. The romance. Add in one of the world's most anticipated aircraft shows, and you have *the* definitive reason to visit. But just make sure you plan your trip around June. You wouldn't want to miss nearly 1900 exhibitors from 40 countries, over 200 aircraft on display, and international aviation appreciation.

Okay, even if you don't plan to visit this electrifying spectacle, get a second-best thrill with past show facts, figures, and photos. Dip into last year's exhibitors' list and aircraft. Read about show history. And, of course, peruse the pictures. Scan the exhibitors for the Concorde and Challenger 604. Get up close to the Boeing 777 and a little Skyhawk. There's Harriers and Hawkeyes, Appaches and Airbus A300s. Pick your favorite alphabet letter and be instantly transported to display aircraft.

What's that? You *are* planning to attend one of the greatest air shows in history? Great! Begin your planning here. Get the costs, the site layout, and general information for visitors. Handy public transportation maps, motorways, telephone numbers, and more are thoughtfully available for the clicking.

If it's supposed to get you excited, actually stirring you into a flying show frenzy, Paris Air Show online succeeds and then some.

Free or Fee: Free.

Europa #272 ZK-TSK, a Builder's Log

http://www.kaon.co.nz/europa

e-mail: none available

RATING
✈✈✈

BRIEFING:

Thorough builder's log puts you smack in the middle of foam, metal, resin, and glass.

The problem I've often tripped over among builder's log Web sites is lack of page updating. Not that I don't understand the concept, mind you. I put building an airplane way ahead of Web site documenting in my time inventory book. Finding someone accomplishing both—building and documenting with meticulous attention to detail and timing—is obviously rare.

So begins our online building adventure with Tony Krzyzewski's Europa #272 ZK-TSK, a Builder's Log. This extensive, almost daily account summarizes the triumphs, tragedies, and tribulations of building an airplane. It's written well. Photos, when included, are appropriate, not gratuitous. And the honesty of this first-time builder makes for fascinating reading. Each annual log (a bit over 5 years worth at review time) is broken down by month. Succinct titles give you a clue as to the log's content: "The Adventure Begins," "The Tail of a Plane," "Winging It (At Last)," "Flapping On," "Control System Rigging," "Cockpit Module," "Lower Fuselage," "Getting Stuck In," and many more monthly entries.

Site navigation is simply scrolling, reading, and moving to the "next" entry. Or use the monthly list of clickable bookmarks. Either way, you're guaranteed great aviation entertainment for your browsing dollar.

Free or Fee: Free.

Dean Garner

http://www.deangarner.com

e-mail: online form

RATING
✈✈✈

BRIEFING:

Visually captivating flight photo site reminds you why you began flying in the first place.

With thousands of badly designed aviation photography and illustration sites crowding cyberspace with bandwidth-hogging files, you may find this site's award mention surprising. But Dean Garner's Impressions museum deserves a look.

The pages are clean, clutter-free, and visually spectacular. Even with gallery after gallery of photos, the site never stalled as my modem hummed along easily. So the mechanics of my experience were good. But the photos really earned the award-winning mention. Quite simply, they're breathtaking. Mostly military, these online gems stop you in your tracks and cause you to pause. At cruise, after takeoff, in vertical climb, in formation, and against scenic backdrops, these aircraft photos remind you of the visual romance of flight.

Each page introduces you to beautiful aircraft-in-flight photos. Just click to enlarge. You'll click into a limitless array of F-18s, A-7s, F-16s, F-4s, F-15s, and even an A-4 or two.

Free or Fee: Free.

Aviation Animation

http://avanimation.avsupport.com

e-mail: jself@accn.org

RATING

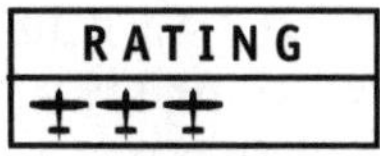

BRIEFING:

Fun and frivolous flying fancies o'plenty.

Purely of entertainment value, this toy box of aviation-related goodies guaranties a smile. It worked for me. With flying aircraft graphics, moving props, and more, how can you go wrong?

An approved archive site from the Animated GIF Artists Guild (lofty credentials, no?), Aviation Animation serves up way more than just animated stuff. Although it doesn't offer up much in page design, its array of clickable fun is endless. Countless animations, aircraft sounds, and movies are scattered everywhere like hidden Easter eggs. Mostly, though, you'll be playing with over 325 aviation animations of antique, military, and present-day aircraft. Rotors spin, props turn, and background scenery goes by. Plan to have snacks and beverages at the ready, because you might be here awhile giggling like a kid.

When you break from animation, try the sounds. The endless WAV files give you file size before loading (thoughtful) and a brief description. Hear running engines, flybys, communications, jet wind, and more.

If you're so inclined, there's even information here on how to make animations. I think I'll just click and enjoy, thank you.

Free or Fee: Free.

Artie the Airplane

http://www.artiebooks.com

e-mail: none provided

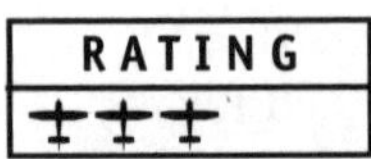

BRIEFING:

A for-kids aviation site that's actually fun for everyone—moms, dads, and future aviators.

Whimsical, fun, friendly, and inviting, Artie the Airplane expands on its popular children's books with a first-rate, just-for-kids Web site. Sure, it's kind of a subtle promo for the books, but you won't mind a bit with this free aviation-rich playground for kids.

Your main stop, with kid in tow, should be the Fun Zone. Here you'll meet Artie the Airplane (a cute, happy-looking rescue plane) and a menu of kid stuff to do. Click into the coloring book for black-and-white drawings to print and color (Jack the Jumbo, Waldo W. Wing, and Wally the Widebody, to name a few). Next, have fun with mazes, connect-the-dots, and maps in the games section. Or meet Artie's friends, complete with great illustrations and descriptions. There's Alice the Air Ambulance, Gramma & Grampa Cubbie, Superslim, and others—all worthy of a look. Finally, as night falls and your mouse hand is tired, read your little flying fan Artie's good-night poem, almost guaranteed to produce happy drowsiness.

All site illustrations are fun and inviting, teamed up with plenty of white space and easy reading for younger Web viewers.

Free or Fee: Free.

SkyFlash

http://www.sky-flash.com
e-mail: ej@skyflash.demon.nl

RATING
✈✈

BRIEFING:

Striking volume and quality launch this flying photo site onto your bookmark list.

I must admit, if it's not already obvious by now, I'm a sucker for flying pictures, no matter the subject. From military to airline to general aviation, quality aviation photos stir me up. Presentation has to be good, and picture quality needs to be above average to capture my viewing time. That's why SkyFlash moved into award-winning status in my opinion.

Mostly originating from air shows, military databases, and display team galleries, SkyFlash pictures fill up your nonflying ground time. Click into the grid/menu of gallery selections or skip directly to the yellow cells identifying the updated stuff. Looking for a preview? Imagine the screaming action of The Snowbirds, Blue Angels, Thunderbirds, high-speed pictures, F-117 Stealth, SR-71 Blackbird, Harrier, Edwards AFB Show, close-ups, helicopters, Team 60, and more.

Links, Special Interest Topics, and Airshow Report News round out your site options, mainly serving as filler.

Free or Fee: Free.

Historic Wings

http://www.historicwings.com
e-mail: none available

RATING

BRIEFING:

This well-rounded presentation explores history in-depth and sports a pretty fancy exterior.

Truly magnificent design draws you in, but the well-written history brings you back for more. Enter the visually spectacular world of Historic Wings.

Brought to you by Capstone Studio, this gem of fun flying remembrances dazzles with such a wide array of Web-savvy solutions. It's online design 101 and then some. Click into the main page just to see the visual theme change. I'm not sure how many "opening looks" there are, but each is well done. Once you delve in, you'll find the simple frames-based navigation easily useful. The efficient menus offer only a few options, steering clear of clutter. "White space" makes online reading and viewing a pleasant experience. Even a Web site translation option offering six (yes, six!) languages rounds out your thoughtful site features.

The wonderfully written historical features are changeable, but the selection and variety are endless. It's quality and quantity! My options (complete with summaries) at review time will give you a preflight inspection on what you can expect: Pretty Deadly (nose art), High Flight (poetry), Flight School 101 (interactive flying inspiration), Learn to Fly Forum, SR-71 Blackbird, B-52 Stratofortress, Aviation Posters (1910s–1940s), and more.

Worthy of one of your top bookmark spots, Historic Wings delivers on style as well as content.

Free or Fee: Free. Subscribe by leaving your name and e-mail for updates.

FlightDeck

http://exn.ca/Flightdeck

e-mail: sysop@exn.ca

RATING

✈✈✈

BRIEFING:

Mostly Canadian historical peek thoroughly delights.

Clean, charismatic, and clearly Canadian, FlightDeck provides a historical flyby courtesy of two respectable industry heavyweights: Discovery Channel Canada and the Canada Aviation Museum.

To thoroughly inspect FlightDeck's aviation museum online, you'll need full multimedia capability. Even if you have to get a free RealPlayer plug-in or download RealVideo, the time logged into FlightDeck is worth it. Step into The Hangar for a look yourself. Get the real-life historical perspective of an Auro CF-105 Arrow, a Bae AV-8A Harrier, a de Havilland DHC-2 Beaver, a Messerschmitt Me 163B-1a Komet, a Submarine Spitfire, and many more. Every aircraft comes to life on your screen with RealVideo clips and brief commentary. See rare archival footage of the planes in flight, or hear some engines revving.

Next, meet some aviators. Skip through a Milestone section of Canadian aviation. Glance into a searchable Image Gallery. Read about the many features and events, or just chat. Whatever your viewing perspective, you'll be visually delighted.

Free or Fee: Free.

The Flight

http://www.gruner.com/flight
e-mail: online form

RATING
✈✈✈

A 50-year-old pilot and a 50-year-old Cessna 195 trip the sky fantastic with this amazing personal account.

It's ironic that such an entertaining site devoted to hearty solo navigation offers slightly awkward Web site navigation. Just slip into the Table of Contents page, and you'll be on course instantly.

Well-written chapters chronicle a 6000-mile journey in a Cessna 195 radial engine beauty. Each online account uses powerful prose and striking photos to capture the essence of one man's fascinating skyward journey. His route includes St. John's, Newfoundland (the most easterly tip of North America), and flying due southwest, over the Canadian maritime provinces, across the United States, through northern Mexico, past the Tropic of Cancer, to Cabo San Lucas, Mexico (the most southwesterly point of the continent). Scintillating chapters are entitled, "Silent Giants of the Atlantic," "Icing, Winds, and Silence," "Flying the Gauges," "Revolutionaries and Bandits," "The Mythical Island of California," "A Small, Dusty Airport," and "The Pearl of Loreto."

This real account speaks of dark fjords, the brutal Sierra Madre, miles of oceans and deserts, and a mixed bag of weather. No, his wife didn't go.

Free or Fee: Free.

Dave English's Great Aviation Quotes

http://www.skygod.com

e-mail: Dave@skygod.com

RATING

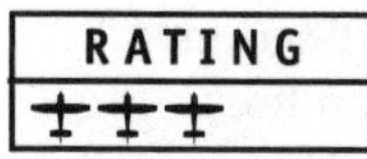

BRIEFING:

Dave English and friends get lofty with words. Soak in this thought-provoking mish-mash of visionary verbiage.

Yes, there's some in jest, but others are rooted in seriousness. However, all the aviation quotes (more than you can imagine) do entertain. The hordes of quotes are conveniently broken down by category. Here's the list (take a breath first): Airports, Air Power, Balloons, Birds, Bums on Seats, Combat, Clichés, First Flights, High Flight, Humor, Last Words, Magic and Wonder of Flight, Stuff, Piloting, Poetry, Predictions, Safety, Space Flight, Women Fly, and ORD ATC.

When you've finished absorbing these thoughts from meandering minds, click through Dave English's other site gems. Learn about Great Drinking Games, Explanation of Airport ABCs, Brain Bags, Just for Fun, Emergency Checklist, and more. Plot a course through tremendous aeronautical chart resources. And as always, stop by the links list—there's plenty.

Free or Fee: Free.

The Mile High Club

http://www.milehighclub.com
e-mail: info@milehighclub.com

RATING

BRIEFING:

A scintillating site devoted to the erotic club that pilots, flight attendants, and daring airline passengers have been whispering about since early flight.

Tastefully exploiting the romantically inclined members of the Mile High Club, the Mile High Club home page boldly makes its online presence known. Mostly text-based, this mile-high adventurer's site organizes a thoughtful mix of limited but appropriate graphics, page navigation perfection, and interesting stories.

Curious Mile High wanna-bes and members will find nonstop entertainment with this erotic fixation on skyward fantasies. Clickable sections include All About the Mile High Club—membership requirements and rules/regulations; Flight Referrals—information on one airline that specializes in this sort of thing; Mile High Club Store; and of course, Tales of the Mile High Club—adult-oriented stories of those in pursuit of becoming a member; and more.

Okay, you're curious, right? Well, here's a taste of the tantalizing tales: "Full Aft Position," "Perks of the Job," "Commuter Fun," "Newlyweds," "Flight to Down Under," and more.

WARNING: Content of Mile High Club stories may contain language or situations not appropriate for minors.

Free or Fee: Free.

Dave, Carey, & Ed's Lancair Super ES Kitplane Progress Page

http://www.edlevine.com/lancair

e-mail: DaveCareyandEd@usa.net

RATING
✈✈

BRIEFING:

Follow along as three first-time kitplane builders take us for a ride on their Lancair learning curve.

By including Dave, Carey and Ed's Lancair Super ES Kitplane Progress Page among the top 500 aviation Web sites, you probably think I've pulled a few too many Gs and rendered myself unconscious. While those around me may disagree, I assure you that I'm fully aware and rational as ever. You'll just need to ride right seat with me on this one and follow my lead.

First, a few warnings. You'll be horrified by site design. Site navigation simply means using your scroll bar. And the long-winded text tends to run on through carefree punctuation. Sweeping all the negatives under the rug, however, reveals an entertaining over-the-shoulder look at the slow progression of building a kitplane. Complete with photos and brutally truthful chronicles of the construction, each entry summarizes the challenges and triumphs.

Although the site doesn't get updated regularly (hey, they've got building to do), you can request automatic e-mail notification when updates occur. When it's completed in a couple of years, the kitplane will evolve into a 220-mpg sport plane. Read on about the Lancair and performance specs by clicking the hyperlink to Lancair.

Free or Fee: Free.

Virtual Horizons

http://ebushpilot.com/horizons.htm
e-mail: Horizons@BushPilot.com

RATING
+++

BRIEFING:

Expand your browsing horizons with an insider's account of the Canadian bush pilot.

Virtual Horizons—it's where the soaring spirits of Canadian bush pilots shower us surfers with cyber-collections of stories, articles, letters, and photographs. Mostly navigating through the Great Canadian North, where even industry-hardened pilots clamor for a window seat, some of Canada's best aviation writers and flyers give you a tour into their experiences.

There's talk and pictures of beautiful landscape and memorable journeys with Cessnas, Beavers, and Otters. Feature stories span a trip into the high Arctic to flying a Cessna Caravan in West Africa. Although giant, high-resolution pictures run wild throughout the descriptive accounts, don't scroll past them unviewed. The download time might be a bit lengthy, but these romantic beauties deserve a peek.

Free or Fee: Free.

Solo Stories

http://www.geocities.com/CapeCanaveral/3831

e-mail: fly@poboxes.com

RATING

✈✈

BRIEFING:

Sensational solo remembrances—*sans* instructor.

The fact that the Solo Stories is easily navigated and packed with stories has nothing to do with my overwhelming recommendation. The brutal honesty and unbridled realism are why I've cleared room for a bookmark.

Granted, Solo Stories brings back personal memories of my own solo and cross-country days, but I'm sure it will do the same for others. Whether you care to relive your solo adventures through the memoirs of others or learn from their mistakes before you fly, check in here for a taste of reality. For a cyber-ride of jittery, real-life flights, I suggest clicking into stories entitled, "If You Run into Traffic, Tell Them You're a Student on Your First Solo—They Will Get Out of Your Way," "As I Latched the Door Behind Him, My First Words Were 'Oh Sh_ _'," and "When I Landed Safely, I Felt Like Amelia Earhart!"

Then, when you're ready to recount your experience, a link guides you into adding your own solo story. It's simple to contribute. Just look at the many entries already showcasing their maiden voyage.

Free or Fee: Free.

Paper Airplanes

http://www.net-www.com/planes.htm

e-mail: webmaster@pchelp.net

RATING

++

BRIEFING:

Perfect for the home, office, or outdoor adventure, these gliders make your childhood spirits soar!

Remember when grandpa expertly added those taped ailerons to your simple paper airplane? Well, I'm sure a lot of aviation wisdom and engineering dreamers came up with this little jewel of a site.

Although this site's focus doesn't necessarily concentrate on anything aviation, the Paper Airplane page it offers just screams, "*Stop what you're doing and have a little fun!*" From an Origami Aerobatic Design to a Supersonic Fast Flyer to a Soaring Glider, anyone running the age spectrum will delight at these propeller-less paper creations. There's plenty of designs to try. Step-by-step diagram instruction glides you through assembly (although instructional pages are a bit fuzzy).

It really is simple, free fun. Click, print, fold, and enjoy!

Free or Fee: Free.

"Dad" Rarey's Sketchbook–Journals of the 379th Fighter

http://www.rarey.com/sites/rareybird

e-mail: dr@rarey.com

RATING

BRIEFING:

An illustratively chronicled tribute to a World War II fighter pilot that deserves a look.

Painful at times and uproarious at others, this wonderful hidden gem of a site delves into a personal account of Mr. George "Dad" Rarey. Drafted into the Army Air Corps in 1942, this young cartoonist and commercial artist kept an animated cartoon journal of the daily life of the fighter pilots. Brought to the bookshelves and now cyberspace by his son and his wife, this thoughtful reflection chronicles "Dad" Rarey's WWII life.

Skillfully prepared and graphically rewarding, this home-page tribute has all the stuff that comes under the heading of great organization: excellent page links, clickable menu icons throughout, and tiny thumbnail pictures that don't waste time (and can grow at your command). But by far the best page features are the written descriptions and illustrations. Contributions you find here come from surviving members of the 379th Fighter Squadron, excerpts from Rarey's letters to his wife (Betty Lou), and Betty Lou's memoirs.

Clickable sections include Cadet Life, Volumes 1–5, Nose Art, and Artifacts.

Free or Fee: Free.

Bookmarkable Listings

Aviation Adventure Stories
http://www.bushwings.com/stories.html
e-mail: none provided
Over 20 individual adventure stories by bush pilot Ron Fox.

World Airshow News
http://www.worldairshownews.com
e-mail: weinman@mailbag.com
Printed trade magazine offers many articles online from current and past issues.

Luc's Photo Hangar
http://www.bayarea.net/~hanger
e-mail: hanger@bayarea.net
Pictorial view of World War II aviation history in the form of aircraft nose art.

Air Pix Aviation Photos
http://www.airpixphoto.com
e-mail: charlie@airpixphoto.com
Collection of aviation photos and related for-sale products.

Miss America Air Racing
http://www.missa.com
e-mail: missateam@aol.com
Get news, event information, aircraft stats, and photo gallery access.

Lost Birds
http://www.lostbirds.com
e-mail: lostbirds@cs.com
Visually spectacular site introduces some historic perspective on aviation mishaps.

Avi8.com
http://www.avi8.com
e-mail: webmaster@avi8.com
Flight forums and religiously updated news.

Delta Flight
http://www.deltaflight.com
e-mail: dave@deltaflight.com
Thorough peek at the fascinating construction of a home-built Boeing 767 flight simulator.

FlightAttendants.org
http://www.flightattendants.org
e-mail: admin@flightattendants.org
Fun, facts, and forums for flight attendants.

Airlinerphotos.com
http://www.airlinerphotos.com
e-mail: photos@airlinerphotos.com
Over 1300 quality airliner photos divided into various thematic sections.

Rotors Helicopters—Hangar Page
http://www.rotors.net
e-mail: gwh@rotors.net
Located in Melbourne, Australia, Rotors serves up some fun puzzles, "HELI-board" message board, and industry news.

Millennium Flight
http://www.millenniumflight.com
e-mail: info@millenniumflight.com
Well-designed site that chronicles the record-breaking flights of Hans Georg Schmid and his Long-Ez.

Airshowreport.com
http://www.airshowreport.com
e-mail: ugamedia@wirehub.nl
Free international air-show magazine by many of the world's finest aviation photographers.

Aviation Employment

JSfirm

http://www.jsfirm.com
e-mail: admin@jsfirm.com

RATING
✈✈

BRIEFING:

Your privacy protection and quality of listings get JSfirm off the ground.

Personal information. Yes, it's floating around Web-wide, and those who choose to remain as anonymous as possible need to be careful. With JSfirm, your personal information is safe. The company takes great pains to protect you, under login names and passwords, and shelter you from the information-grabbing Web hounds. Specifically developed to attract both employers and employees together without divulging confidential information, JSfirm's online aviation employment stopover gives you the best of both worlds—quality in aviation employment matchmaking and privacy protection.

With its modestly designed Web site (obviously built for ease of use and functionality), JSfirm displays its employment wares with simplicity. Are you searching for aviation employees? The site helps aviation companies locate qualified aviation employees in two ways: as a resumé source or hire to help locate aviation employees. The site deals in such positions as avionics technicians, A&P mechanics, aircraft maintenance technicians, sales, aerospace engineering, project management, executive, finance, accounting, pilots, and more.

Listing-wise, the resumé and available jobs pool was a little low at review time. But with an obvious attention to quality and privacy concerns, JSfirm is bound to soar.

Free or Fee: Free to employees. Fee for employers.

AirlineCareer.com

http://www.airlinecareer.com

e-mail: support@airlinecareer.com

RATING

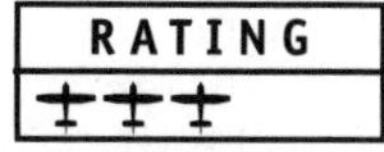

BRIEFING:

Flight attendants and wanna-bes need only stop in here for career information.

Probably my favorite feature of AirlineCareer.com (and there are many) is simply the site's specific employment focus on flight attendants. Not every aviation occupation known to humanity. Just flight attendants. So it should come as no surprise that what lies within AirlineCareer.com, geared only to the flight attendant and wanna-be, is valuable industry information. And as you may expect, it's not free. There is a fee to use AirlineCareer.com. But industry insight this juicy demands a little give and take. Just how juicy, you ask?

Well, after you've received your user name and password (fee-related), delve into the very researched world of AirlineCareer.com. A comprehensive Evaluation Center begins your journey and helps to evaluate your new career (pay, hours, duties, tips on getting hired, etc.). The Application Center then helps you choose the right airlines, prepare your resumé, fill out applications, and more. Sixty real interview questions with answers get you completely prepared. And a well-researched section on hard-to-find airline pay rates for 10 major airlines gets to the heart of the matter.

AirlineCareer.com pushes its cart down the isle with other needed tidbits too, such as interactive message board, flight attendant dictionary, easy-to-use search engine, industry links, and more.

Free or Fee: Nominal fee to use service.

Climbto350.com

http://www.climbto350.com

e-mail: admin@climbto350.com

RATING

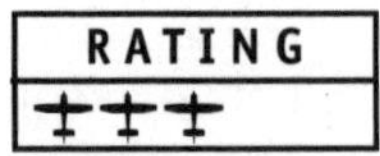

BRIEFING:

Innovative aviation jobs board throws wireless access into the mix.

This huge, if slightly unwieldy jumbo jet of a job board lumbers into cyber airspace with a steady climb. From takeoff, instructions are clear, controls are well positioned, and instructions are plentiful. Rest easy, you'll enjoy your ride with Climbto350.com.

What's most cool about Climbto350.com isn't necessarily the enormous (and growing) number of posts. Rather, it's the fact that Climbto350.com uses the latest in wireless communications technology to provide aviation professionals, flight attendants, mechanics, flight engineers, dispatchers, pilots, and others with current job information.

Accessible anywhere in the world by Internet, cellular laptop computer, or personal Web phone, this innovative aviation jobs board gets to the heart of the board quickly. If you happen to be that wireless techie who has the latest in surfing technology, you will need to know a few things first. To access the Wireless Worldwide Aviation Jobs Board at Climbto350.com you'll need a cell phone with Web browsing capability. You'll also need to access the different, wireless-specific URL of http://www.climbto350.com/wap/js.wml. This URL address only works with cell phones that have Web browsing capability. Cool, huh?

Free or Fee: Free.

PlaneJobs.com

http://www.planejobs.com

e-mail: customerservice@planejobs.com

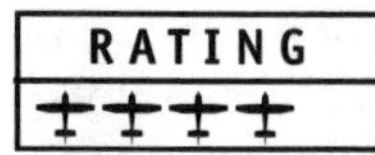

BRIEFING:

Well-crafted take on how aviation employment online should go.

PlaneJobs.com gets it. The company understands perfectly how to use the Web to bring aviation employees and employers together. Keeping things simple, clean, and quick, PlaneJobs.com uses little in the way of fancy graphics (although minimalist layout and design are top-notch) and concentrates on its reason for being—jobs.

Founded in 1999, the company continues to increase its listings and industry awareness. As of review time, some of the job categories contained a less than stellar quantity, but watch for the growth. Don't get me wrong, there were certainly enough quality job listings to warrant a careful review. Just remember to bookmark it, though, for additions.

The scope of aviation job categories demonstrated a nice cross section of the industry. See if you agree. Currently clickable areas are accounting and finance, engineering, flight operations, ground operations, human resources and training, information technology, maintenance, management, pilots, planning and scheduling, purchasing, properties, and facilities, sales and marketing, and reservations/customer service.

Free or Fee: Job seeking is free. Employer services are fee-related.

PilotsWanted.com

http://www.pilotswanted.com

e-mail: online form

RATING

BRIEFING:

Information wanted? Career pilot jobs wanted? Go to PilotsWanted.com.

The Berliner Aviation Group, the company behind PilotsWanted.com, is made up of over 25 airline pilots and aviation professionals dedicated to helping match clients with airline and corporate pilot jobs. All staff members fly as line pilots and check airmen for some of the largest carriers in the country. Did that get your attention? I thought so.

Okay, so where's the list of jobs, you ask? Not so fast. PilotsWanted.com focuses on one-on-one services and information rather than on laundry lists and mass employment. Lots of thorough product/service descriptions promote the company's personalized resources in Book Recommendations, Career Pilot Job Reports, a Resumé Referral Service, Resumé Evaluation, Practice Interview Sessions, and more.

If you're just transitioning through, however, looking for some quality career pilot insight, PilotsWanted.com does that too. Well-written FAQs dealing with career recommendations and tactics are excellent. The latest job listings are a click away (membership required). And of course, there's the many years of information-packed online newsletters to review—guarantying at least an afternoon of learning.

Free or Fee: Lots of free information. More expanded services and products are fee-based.

The CoPilot.com

http://www.thecopilot.com

e-mail: feedback@thecopilot.com

RATING
✈✈

BRIEFING:

Give The CoPilot.com a little time, and watch it move into the industry's left seat.

With The CoPilot.com I'm gambling a bit. I'm betting that with time (maybe by the time you read this) this aviation jobs site will catch fire and the quantity of listings will explode exponentially. Why? Because the site mechanics are in place. The database works well enough. The layout's functional. And everything is free! Yes, no matter what side of the aviation employment coin you're on, The CoPilot.com offers up its services for both employers and potential employees for *free*. No setup fees, no monthly charges, and no membership dues to pay.

Position seekers will enjoy the quick-access database offering a table view of jobs, followed by a detail page of job specifics. Get at-a-glance facts on position type, company, location, wage or salary range, description, and minimum qualifications. Pilot searching serves up specific pilot resumés and details such as flight time, ratings, licenses, experience, desired position, etc.

It's worthy of note that other things, beyond job stuff, fly around site-wide in uncontrolled airspace, too. Weather information, flight planning links, shopping (mainly aircraft models and books), and want ads (sparsely populated) fill unobtrusive menus. My advice: Stick to the job stuff.

Free or Fee: Free.

JetCareers

http://www.jetcareers.com
e-mail: doug@jetcareers.com

RATING
✈✈✈

BRIEFING:

Honest and informational, this little gem of knowledge glistens in the murky sea of airline hiring.

With little fanfare or big-dollar Web design teams, this gritty little site tucked away in cyberspace rates high with wanna-be airline pilots looking for honesty. Succinctly summed up by the illustrious site author, "There are plenty of commercial sources of information about careers in aviation; however, I saw a need for an unbiased source of information made by professional pilots for future professional pilots and other people that are interested in a 'behind the scenes' perspective."

Religiously updated and brimming with employment tips, techniques, and down-to-earth advice, JetCareers steers you into many areas of interest. The Main section welcomes you with the JetCareer Newsletter, airline industry news, discussion groups, and a chat room. In Education, you'll find information on college degrees, choosing a flight school, the civilian route, the military route, and more. Airline Flying talks about new hire training, how your site author reached the cockpit, frequently asked questions, and "A Day in the Life." Sections involving corporate aviation and job hunting round out your site options.

After much quality reading and learning at JetCareers, I recommend a layover at the Cool Stuff section. Here, you'll uncover "Pictures from the Road," aviation links, a live link to Dallas Fort Worth Tower, and the ability to track any flight.

Free or Fee: Free.

AvEmp.com

http://www.avemp.com

e-mail: tech_resources@avemp.com

BRIEFING:

Information design at AvEmp.com propels this aviation match-maker into award-winning status.

Free or Fee: Everything for pilots *and* employers is free (except the ad banners).

Relatively new on the cyber-scene at review time, AvEmp.com throws its hat into the online aviation career ring. It has joined the fray with pilot job listings, resumés online, and discussion groups. A few links, an e-mailed employment information service, and a handy "site progress" are also available during your visit.

The beauty with AvEmp.com is details. The searchable information presents itself well. And as I'm sure you've noticed, not every aviation employment site knows how to display its jobs, pilots, and resumés effectively. Specifically, the input forms and resulting displays show expert organization. The left-seat database pilot behind the scenes here knows his or her stuff. Try to search or add information yourself. It's effortless and intuitive. Don't expect any downtime either. Because everything is text-based (menus and information display), AvEmp.com satisfies your need for speed.

The completely free options (for pilots and employers) fill the left-margin menu as Pilot Resumés (list new resumés, list all resumés, and search resumés), Aviation Jobs (list new jobs, list all jobs, and search for a job), Discussions (many topics, few posts as of review time), and Help (lots of excellent, must-read FAQs).

Aviationjobsearch.com

http://www.aviationjobsearch.com
e-mail: sales@aviationjobsearch.com

RATING
✈✈

BRIEFING:

Worldwide pilot search and posting site finds itself high atop the bookmarks list.

Mostly United Kingdom–based, or at least European-focused, Aviationjobsearch.com dips into a huge pool of job openings (well over 6000 at review time) and presents them well to interested viewers. You'd think with so many listings and obvious popularity, Aviationjobsearch.com would grind to a crawl as you summons the database behind the scenes. Not so. Snappy and pleasing to the eye, the data pop up at your command. Your left-margin choices include Joblist E-mail (see description of e-mail notification service below), Vacancies by Job, Vacancies by Employer, Agency Vacancies, Courses & Training (summarized company information and offerings), and more.

A feature worth mentioning, although not new by any means to aviation employment online, is the e-mail delivery notification of new jobs that match your specific requirements. This free list of vacancies arrives at the e-mail address of your choice daily. Now that's service.

Free or Fee: Free to view openings, fee for employers to advertise jobs.

AviationCareer.net

http://www.aviationcareer.net
e-mail: editor@aviationcareer.net

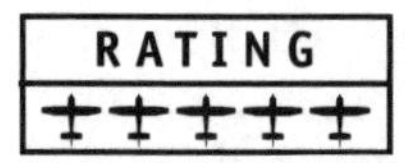

BRIEFING:
Well-designed aviation career magazine headed on a collision course with greatness.

Well, prepare for landing, job seekers, we've arrived at our final destination of aviation career information perfection. Not sure if it gets better than this. May as well stow your tray table, cinch up the seat belt, and add this one to your favorites list.

AviationCareer.net exists as a virtually flawless online "zine" directed toward aviation career and employment issues. The primary browsers are employees, job seekers, management, and human resources professionals in the aviation industry. In typical magazine style format, AviationCareer.net displays its wares in the hovering left-margin menu, with corresponding content filling the main screen. Your options for employment enrichment are many. Reserve the better part of a rainy day sifting through Ask the Expert, Fly By (a weekly column of tidbits, news, and oddities in the aviation industry), Job Hunting Tips, Training Department, and Recommended Resources (jobs, company search, aviation career airfairs, chat lounge, links, bulletin board, calculators, and more). If this isn't enough, be sure to read the many feature articles. These constantly updated articles are reason enough for a bookmark.

Free or Fee: Free.

Direct Personnel

http://www.directpersonnel.com

e-mail: dpi@iol.ie

RATING
✈✈

BRIEFING:

Aviation recruitment specialist in Ireland attracts a worldwide audience.

I'll admit that Direct Personnel may be a stretch for this book. Its focus is narrow, primarily serving as an aviation recruitment tool for high-end international pilot and maintenance jobs. Although you may not fall into this category, you might know someone who does.

No, it's not brimming with quantities of jobs (at time of review), but every resource helps, right? Read through the company background, goals, benefits, and update notices of jobs added to the database. Eager to get a taste of jobs available through Direct Personnel? Just to give you an idea, here's the partial list compiled at review time: B747 First Officers, Bae 146 Captains and First Officers, Captain IRE/TRE A330, MD 11 First Officers, and Saab 2000 Captains. Maintenance jobs also can be had at Direct Personnel. Again, here's just a slice of the positions available: A310-300 Engineer, Fokker 50 engineers, and MD-11 Licensed Engineers.

Free or Fee: Free to rummage around in the site.

Universal Pilot Application Service (UPAS)

http://www.upas.com

e-mail: support@upas.com

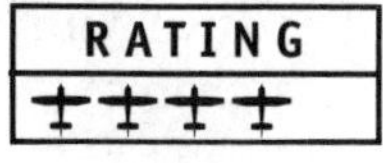

BRIEFING:

Well crafted, UPAS online demonstrates Web perfection in matching pilots with jobs.

The Universal Pilot Application Service shows off its Web skills with a thoroughly captivating aviation employment megasite. UPAS, an aviation employment powerhouse, uses an innovative approach in matching pilots with companies. Although the service is fee-related, you'll need to check into UPAS to fully realize the potential here.

By completing the comprehensive Qualifications Summary, a pilot has the opportunity to provide an extensive preview of his or her flying experience with would-be suitors. At review time, UPAS had more than 7000 active pilots in the database. Flight-experience levels vary from single-engine flight instructors to Boeing 747/400 captains.

Companies checking into UPAS to review the talent pool include representatives from global airlines, regional airlines, training centers, commuter airlines, international and domestic corporations, crew leasing organizations, and cargo and charter operations.

Free or Fee: Fee-related services.

AvCrew.com

http://www.avcrew.com

e-mail: info@avcrew.com

BRIEFING:

With completely current jobs posting, AvCrew.com is always on time.

Yet another crew employment site takes the active with this outstanding online service. Succinct, searchable, and simple to navigate, AvCrew delivers on its mission: "to present more employment opportunities to pilots and give employers a qualified applicant pool, faster than traditional methods."

Employers searching the skies for left as well as right seat applicants will find a growing collection of pilot resumés, positions wanted, and classifieds. Faster than you can say FlightSaftey trained, a search for qualified corporate pilots begins and ends with AvCrew. Employers can post jobs (fee-related), browse pilot resumés, browse other crew member resumés, and search for qualified applicants. Site resources equally cater to pilots and employers. Pilots searching for dream assignments need only scan the current positions list and click into more detailed description. Position details and contact information ease you into a stress-free search. Really. With no distractions from ad banners, giant pictures, or audio gizmos, AvCrew gives you easy viewing for position searching.

To summarize, many nominal-fee extras combine with quality free stuff to create a solid source for aviation employment. Whether you go fee or free, fly with AvCrew.

Free or Fee:
Mostly free. Some services have nominal fees.

Aviation Employment.com

http://www.aviationemployment.com
e-mail: info@aviationemployment.com

RATING

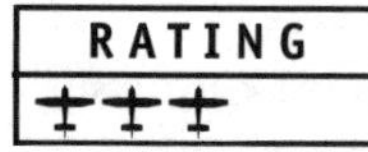

BRIEFING:

Worldwide aviation employment opportunities on an online silver platter.

For years, Employment Publications, Inc., has been doing one thing well: printed aviation/aerospace employment guides. Today, the company continues with printed excellence in the form of *Pilot Employment News* and *Aviation Maintenance & Engineering Journal*. Now the company has expanded its successful single employment vision to the Web world. Sure, it's a natural transition, but a satisfying compilation of page elements isn't so natural. In fact, it's elusive to most digital dabblers. But Aviation Employment.com has managed to rise above the mediocrity with its clean, inviting, and easy-to-use online resource.

Your seven menu options that follow you to each page are Job Listings (lists an alphabetized laundry list of positions by title), Companies Listed, Employment Guides, Submit Resumé, Schools & Training, and About Us (get to know the company behind the coding).

As you'd expect from pros, there are many job listings to review, so a well-designed search engine is fueled and at the ready. Just identify your requirements and qualifications to find the matches. Narrow selection by state, your current educational level, type of aviation job you seek, full- and/or part-time positions, permanent and/or temporary, and salary requirements.

Free or Fee: Free to browse, but fees apply to those wanting to post jobs.

Air Base

http://www.airforce.com
e-mail: online form

BRIEFING:

Fancy online Air Force recruiter informs and invigorates potential members.

Straight out of a virtual dogfight scene from your favorite flight simulator game, Air Base cranks up your adrenaline a notch with a heads-up display interface that's second to none. The Air Force recruitment information housed in this online hangar presents itself in a compelling, almost electric way.

Just before you think this site's all sizzle and no beef, don your flight suit and take a ride. It's packed with details and overviews concerning an Air Force career. Yes, it's a pixilated pitch. But if you're contemplating Air Force possibilities, research is merely a click or two away. Here's the lineup: Air Force Live Events serves up cool interactive "Webcasts"; Stealth Force offers animated series with games, downloads, and more; and Air Force Education promotes the wealth of training and education opportunities available to USAF members. Air Force Careers, Health Center, The Air Force Experience, and Information Center round out a full flight line of informational resources.

Free or Fee: Free.

Fltops.com

http://www.fltops.com

e-mail: webmaster@fltops.com

RATING

+++

BRIEFING:

Working behind the scenes, Fltops.com is your personal airline crew job informant officer.

Quietly discussed in cockpits, pilot lounges, and personnel offices throught the industry are hints, tips, tricks, and essentials for landing a flight officer position with a major carrier. Fltops.com owners and editors behind the pages are active pilots with major carriers and continually provide the "Internet-based intelligence" helpful in your professional pilot quest.

Just a look at the introductory page and you'll agree that you've stumbled on a major resource. Jumping out at you are scrolling banners of breaking industry news, top stories, and updates. Get up-to-the-minute insight on such topics as new hourly pay rates for crew members, benefits comparisons among the "big six" airlines, and contact information on the "big 13."

It should be clearly noted that everything on the site isn't free. In fact, Fltops.com is primarily fee-oriented for the good stuff. Become a "crew member" and receive quality industry intelligence. Even if you're just dabbling in Fltops.com's resources, be sure to scan the free section available to all: FlightLine News, airline financial updates, fleet profiles, and a special report for older pilots.

Free or Fee:
Some info is free; more detailed stuff requires fee-related membership.

Aviation Employee Placement Service (AEPS)

http://www.aeps.com

e-mail: aeps@aeps.com

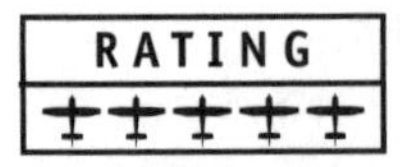

BRIEFING:

Get yourself in front of almost 7000 aviation companies with a few clicks of the mouse.

First, let's talk numbers. As of review time, AEPS boasted over 95,000 potential applicants worldwide, and this number obviously will grow daily. Next, you have almost 7000 companies using AEPS (again, as of review time). Finally, if you're an aviation job seeker, 10 is the last number you need to know—as in 10 free days to sample AEPS. Okay, now bookmark AEPS, and read on.

Employers will find AEPS a thrilling employee resource. It's free to you, always. You'll enjoy free job posting on the job pages, unlimited search capability at your convenience, free preemployment screening checks and compliance in the United States with the Pilot Record Act of 1996, and your own Jobs & Application page.

For everybody, aviation news, career fair information, career links, and help screens round out your AEPS experience. And it is quite an experience. One you'll just need to see to believe.

Free or Fee: Job seekers can try it free for 10 days. As always, job postings and data bank searching are free for all employers.

Air, Inc.–The Airline Pilot Career Specialists

http://www.airapps.com

e-mail: airinfo@airapps.com

BRIEFING:

Plan a thorough career course here before you go wheels up.

To begin, you'll need to make only one choice: applicant or employer. Whether you follow the AirApps path for applicants or the AirAces for employers, you'll roll down the active to superb employment-driven resources. For airline career seekers, Air, Inc.'s lofty online resources span the industry to give you a jumbo-sized heads-up. You'll climb aboard an unrivaled career guide for pilots. Employers will find unmatched assistance in searching for applicants, selecting applicants for interviews, and more.

Once you've committed to the AirApps or AirAces side of things, jump into the online tour for great insight. Applicants will have the opportunity to fill out an application directly online that becomes immediately viewable by the airlines. Make updates as often as you wish with no extra charge. Airline employers will find an equally impressive tour and stockpile of resources. You'll have direct access to pilot applications (searchable, of course), resumé/application screening and telephone interview by Air, Inc.'s aviation specialists, and cognitive computer-based testing for pilots, flight attendants, and mechanics.

Free or Fee: Memberships have many different levels and are fee-oriented.

Once job seekers have committed to membership, the jetway pulls up to religiously updated Hot Air News, U.S. Salary Survey, Pilot Resource Guide, Career Decisions Guide, Address & Information Directory, Fleet & Sim Guide, Searchable Airline Database, the Application Handbook, and more.

Aviation Jobs Online

http://www.aviationjobsonline.com

e-mail: info@aviationjobsonline.com

RATING

+++

BRIEFING:

Award-winning seeker site. Yes, it's fee-oriented, but it does all the grueling work.

Searching for that dream job sometimes requires as many allies as one can muster. Sign up with Aviation Jobs Online and you'll instantly have the beginnings of a powerful job search—24 hours a day, 7 days a week.

Self-proclaimed as "the aviation world's most complete employment source," Aviation Jobs Online is your one-stop employment shop. Although site navigation leaves a bit to be desired (plan on lots of scrolling), informative topics abound. There are free and membership-only areas, special offers and contests, a book store, Post a Job Listing, View Resumés, Post a Position Wanted, View Positions Wanted, Aviation News Now, free job posting for employees, and more.

After you get through the hard-sell areas, you'll find that these folks are serious about aviation jobs with their resourceful personnel service. They'll help with resumés, finding articles on specific companies, and help to prepare you for the job interview.

Over 2000 links, Crew Room Bulletin Board, and Electronic Post Office round out the site's cool creations.

Free or Fee: Fee-related. Many fee options are available—see site for details.

Find A Pilot

http://www.findapilot.com
e-mail: webmaster@findapilot.com

RATING
✈✈✈

BRIEFING:

Meet the no-frills yet focused employment matchmakers at Find A Pilot.

With a few years of success under its Web belt, Find A Pilot takes the active with its own cyber-version of online employment exchanges. The site seems to avoid any gratuitous visual pleasantries—skipping right to the meat of the matter. It's all about jobs and aviation professionals. Period. No dreary news. No QuickTime flybys or audio oddities. Serious aerohunters will appreciate the fluff-free focus. It's refreshing.

A similar text-based menu springs up everywhere, giving even frantic job searchers easy maneuverability. The occupation-only offerings for position-shopping pilots include View Jobs area (free), Post Resumés (fee), FAQs, and more. Personal "homepage resumés" include your choice of background colors or wallpaper, multiple category listings, user-defined links, and free unlimited updates!

Employers shopping for new recruits will enjoy free viewing of the resumé database (a healthy grouping as of review time) sorted in no particular order. Actual job listings for employment seekers are carefully arranged by Corporate, CFI, Helicopter, Part 135, A&P Mechanics, and Miscellaneous.

Free or Fee: Fee-oriented services available.

Airline Employment Assistance Corps (AEAC)

http://www.aeac.net

e-mail: info@aeac.com

BRIEFING:

A resourceful fee-oriented aviation career counselor. You'll find (or fill) that long-awaited aviation position with help from the AEAC people.

You've hung out long enough down at your local airport. Face it, you'll need to get a little more serious if you're going to find an aviation job. However, hold onto your headset, the Airline Employment Assistance Corps is your new online resource. Although the name would imply airline only, there's room here for any career in aviation.

This lofty employment service provides a long list of helpful categories: worldwide classified ads, resumé resources (post your own here, or get resumé help from pros), an industry look at opportunities (titles, salary ranges, education requirements, and employers), airport careers, aviation and maintenance careers, air traffic controllers, aviation safety inspectors, flight attendant careers, government aviation careers, pilots and flight engineers, a salary relocation calculator, and a host of aviation-related links.

Because this is a professionally maintained employment site, a membership fee is required. If you're really an aviation job seeker, the value here is obvious.

Free or Fee: Some free areas, but membership gives you access to everything.

Aviation/Aerospace Jobs Page (NationJob Network)

http://www.nationjob.com/aviation

e-mail: none available

RATING

BRIEFING:

Looking for aviation employment? This *free* professional service does the job.

Offered up by the huge employment resource NationJob Network, aviation opportunities abound here. Access an endless sea of jobs one of two ways: Either click on home-page logo icons for some big-name industry leaders (Boeing, Learjet, Raytheon Aircraft, Cessna, BFGoodrich Aerospace, Bombardier) or search through jobs listed here by location, salary, and more. Simply click on an appealing job in the lineup. From there, you gain access to a company profile as well as an enlightening job description. Most likely you'll be overwhelmed by the variety of categories. Positions range from flight test engineers to buyers and from A&P service mechanics to vice presidents of operation.

Even if you're not thrilled with the arduous task of sifting through these nationwide listings, just ask "P. J. Scout" to do it for you automatically. This convenient little feature makes employment hunting effortless with an e-mail notification service. Simply enter your job preferences and e-mail address. "P. J. Scout" will search the furthest reaches of the Web and find jobs that match your parameters. He reports to you by e-mail weekly. It's free, confidential, and cool.

Free or Fee: Free.

Bookmarkable Listings

FAA Aviation Education—Resource Library
http://www.faa.gov/education/resource.htm#career
e-mail: see site for e-mail list
Downloadable documents with insightful employment information.

Av Canada
http://www.syz.com/avcanada
e-mail: avcanada@syz.com
International aviation employment services.

Corporate Pilot
http://www.corporatepilot.com
e-mail: service@corporatepilot.com
Matching corporate aviation flight departments with pilots and mechanics.

FlyJetsNow.com
http://www.flyjetsnow.com
e-mail: info@flyjetsnow.com
Home of the Airline Career Checklist and self-described as "the only step-by-step path to a major airline career."

Airport Job Hub
http://www.fly.to/airportjobs
e-mail: none available
Free job networking within the airport industry.

Airline People
http://www.airlinepeople.com
e-mail: davidc@airlinepeople.com
Worldwide resource for airline jobs.

AviationJob.com
http://www.aviationjob.com
e-mail: staff@aviationjob.com
Free job listing for aviation employers looking for help.

JobsInAviation.com
http://www.jobsinaviation.com
e-mail: info@jobsinaviation.com
Fee-based online employment service promoting job opportunities for pilots, flight attendants, mechanics, and flight instructors.

Job Air
http://www.jobair.com
e-mail: jobair@compuserve.com
Netherlands-based Job Air recruits, selects, and places aviation personnel worldwide.

Index

About the Author

Merging his 11-year private pilot experience with almost 13 years of corporate marketing management and 5 years of Web site design, John Merry owns Specialized Marketing Agency—a Web design and online marketing company.

In addition to authoring three prior editions of this book, Mr. Merry has written many Web-related aviation articles for such publications as *Plane & Pilot* magazine, *Inflight USA*, and *Plane & Pilot News*. Memberships include the Pilots International Association, Aviation Owners and Pilots Association, Web Marketing Association, the HTML Writers Guild, and the International Webmasters Association.